Measures of
Interobserver
Agreement
and Reliability
Second Edition

Chapman & Hall/CRC Biostatistics Series

Chapman & Hall/CRC Biostatistics Series

Chapman & Hall/CRC Biostatistics Series

Measures of Interobserver Agreement and Reliability

Second Edition

Mohamed M. Shoukri

King Faisal Specialist Hospital & Research Center
Riyadh, Saudi Arabia

Department of Epidemiology and Biostatistics
Schulich School of Medicine and Dentistry
University of Western Ontario
London, Ontario, Canada

CRC Press
Taylor & Francis Group
Boca Raton London New York

CRC Press is an imprint of the
Taylor & Francis Group, an **informa** business

A CHAPMAN & HALL BOOK

CRC Press
Taylor & Francis Group
6000 Broken Sound Parkway NW, Suite 300
Boca Raton, FL 33487-2742

First issued in paperback 2020

© 2011 by Taylor and Francis Group, LLC
CRC Press is an imprint of Taylor & Francis Group, an Informa business

No claim to original U.S. Government works

ISBN-13: 978-0-367-57706-3 (pbk)
ISBN-13: 978-1-4398-1080-4 (hbk)

Library of Congress Cataloging-in-Publication Data

Shoukri, M. M. (Mohamed M.)
    Measures of interobserver agreement and reliability / Mohamed M. Shoukri. -- 2nd ed.
       p. cm. -- (Chapman & Hall/CRC biostatistics series)
    Originally published: Measures of interobserver agreement / Mohamed M. Shoukri. 2004.
    Summary: "From the First Edition:Agreement among at least two evaluators is an issue of prime importance to statisticians, clinicians, epidemiologists, psychologists, and many other scientists. Measuring interobserver agreement is a method used to evaluate inconsistencies in findings from different evaluators who collect the same or similar information. Highlighting applications over theory, Measure of Interobserver Agreement provides a comprehensive survey of this method and includes standards and directions on how to run sound reliability and agreement studies in clinical settings and other types of investigations. The author clearly explains how to reduce measurement error, presents numerous practical examples of the interobserver agreement approach, and emphasizes measures of agreement among raters for categorical assessments. The models and methods are considered in two different but closely related contexts: 1) assessing agreement among several raters where the response variable is continuous and 2) where there is a prior decision by the investigators to use categorical scales to judge the subjects enrolled in the study. While the author thoroughly discusses the practical and theoretical issues of case 1, a major portion of this book is devoted to case 2. He explores issues such as two raters randomly judging a group of subjects, interrater bias and its connection to marginal homogeneity, and statistical issues in determining sample size. Statistical analysis of real and hypothetical datasets are presented to demonstrate the various applications of the models in repeatability and validation studies. To help with problem solving, the monograph includes SAS code, both within the book and on the CRC Web site. The author presents information with the right amount mathematical details, making this a cohesive book that reflects new research and the latest developments in the field"-- Provided by publisher.
    Includes bibliographical references and index.
    ISBN 978-1-4398-1080-4 (hardback)
    Diagnostic errors. 2. Medical laboratory technology--Statistical methods. I. Shoukri, M. M. (Mohamed M.) Measures of interobserver agreement. II. Title. III. Series.

    RC71.3.S478 2010
    616.07'5--dc22
                                2010044854

Visit the Taylor & Francis Web site at
http://www.taylorandfrancis.com

and the CRC Press Web site at
http://www.crcpress.com

I would like to dedicate this work to:

My wife Suhair and my children Nader, Nadene, and Tamer

Contents

Supplementary Resources Disclaimer

Additional resources were previously made available for this title on CD. However, as CD has become a less accessible format, all resources have been moved to a more convenient online download option.

You can find these resources available here: www.routledge.com/9781439810804

Please note: Where this title mentions the associated disc, please use the downloadable resources instead.

Preface to the Second Edition

Since the publication of the first edition of *Measures of Interobserver Agreement* (2004) by Shoukri, three books on the subject matter have appeared on the market. These are the second edition of *Statistical Evaluation of Measurement Errors* (2004) by G. Dunn, *Analyzing Rater's Agreement* (2005) by A. von Eye and Eun Young Mun, and *Bayesian Methods for Measures of Agreement* (2009) by L.D. Broemeling. Our second edition includes new material embedded in two additional chapters that are not in the first edition. One chapter is devoted to discussing various models for methods comparison studies, and the other is for the analysis of reproducibility using the within-subjects coefficient of variation. We have eliminated materials on the patterns of agreement and analysis under log-linear models because they are extensively covered in the book by von Eye and Mun. Besides, we have made no attempt to discuss Bayesian methods as the book by Broemeling would be the reference of choice. The other four chapters are devoted to the analysis and interpretation of results from reliability and agreement studies.

Several issues are emphasized throughout the second edition. The most important issues related to the design and analysis of reliability and agreement studies are the definition of the subjects' and the raters' population. As in several parts of the book, classification of subjects and measuring devices depends on the field of application. Typically, the rater population is characterized by raters' skills, clinical background, and expertise. All these factors affect the degree of measurement errors, and their quantification is needed in reliability generalization studies. In the cases of categorizing individuals into classes, the number of classes is usually predetermined. Reviewing the work by Donner and Bartfay (2002), it will be shown that the level of agreement depends on the number of categories. In the case of continuous measurements where the intraclass correlation coefficients are the most recognized measures of reliability, Muller and Buttner (1994) and Rousson et al. (2002) showed that the values of these measures depend on their range. It is then natural to demonstrate and quantify the loss in efficiency in estimating agreement when a continuous variable is categorized. One of the most important issues in reliability studies is the sample size determination. Depending on the nature of the study, sample size refers to the number of subjects, the number of raters, and the number of replications taken by each rater. At the end of every chapter, sample size requirements relevant to the topic being covered will be discussed extensively.

Information about subjects and raters may be incorporated in reliability studies. Subjects' age, health status, and other demographic characteristics may influence the process of diagnosing each subject—hence, the assessment result. Through logistic regression models, one may account for such sources of variation to accurately measure the reliability of the measuring instrument.

An important fact is that estimation and interpretation of the results of reliability and agreement studies are valid only if the assessments made by raters are blinded. For example, if several interviewers are rating a group of subjects to select a subgroup for a specific job, then ratings made by an interviewer, if not blinded, may influence the other interviewers' ratings. This dependence of raters' scoring will bias the index of agreement or reliability. Blinding and random selection of subjects—and, in some cases, raters—are fundamental assumptions that we need to emphasize at the outset.

In many cases, reliability studies are conducted in several centers, and one of the study objectives would be to produce one pooled estimate of agreement. In this case, we discuss two ways of pooling, the first being the fixed-effect approach, while the second is regarding the multicenters as a random sample of centers, and we use a simple meta-analytic approach to find an overall estimate that accounts for the between-centers heterogeneity.

Several exercises are given at the end of every chapter, and a CD that contains all the data sets accompanies the book. We also provide the reader with the SAS codes for the data analyses.

MATLAB® is a registered trademark of The MathWorks, Inc. For product information, please contact:

The MathWorks, Inc.
3 Apple Hill Drive
Natick, MA 01760-2098 USA
Tel: 508 647 7000
Fax: 508-647-7001
E-mail: info@mathworks.com
Web: www.mathworks.com

Preface to the First Edition

Agreement between two or more measuring devices is an issue of prime importance to statisticians, clinicians, epidemiologists, psychologists, and many other scientists. For example, the recorded readings made during clinical evaluations need to be consistent, whether recorded by the same physician on different visits or by different physicians. Imaging techniques such as magnetic resonance imaging (MRI) and ultrasonography are used to assess the staging of cancer. Such techniques should produce accurate results to ensure efficient delivery of appropriate treatment to cancer patients. Indices to quantify the reproducibility and consistency of readings are the subjects of this book. Since the publication of Graham Dunn's (1989) book titled *Design and Analysis of Reliability Studies: The Statistical Evaluation of Measurement Errors*, there has been a flurry of research in the area of agreement, particularly when subjects are assessed on a categorical scale. The level of complexity of this research varies from simple and practical, to highly sophisticated and beyond routine implementation by the ordinary practitioner.

My motivation for writing this book is based neither on the feeling that I can do a better job presenting known material, nor on the knowledge that I have something new to contribute to the literature. In the analysis of Interobserver Agreement, my presentation of the material focuses on the basics and the practical aspects of the topic. In my opinion, however, heavy mathematical details may not be necessary and should be omitted.

It is assumed that the readers have had prior exposure to the topics of linear regression, correlation, analysis of variance, and cross-classification tables. Moreover, familiarity with the SAS Program is required.

The ideas in the text are illustrated by examples, some of which merely illustrate simple points while others provide relatively advanced data analyses, such as logistic regression and log-linear models. The primary purpose of the book is to provide a practical approach to the analysis of agreement and reliability studies for applied statisticians, clinicians, and scientists.

Acknowledgments

The author would like to acknowledge the support for this work by the Research Center Administration of the King Faisal Specialist Hospital and Research Center. Special appreciation and thanks go to Syed Mohamad Shajahan, assistant manager of Techset for his meticulous work during the preparation stage of this book.

Author

Professor Mohamed M. Shoukri received master's and a PhD in statistics from the Department of Mathematics and Statistics, University of Calgary, Alberta, Canada. He taught applied and mathematical statistics at Simon Fraser University, the University of British Columbia, and the University of Windsor. Between 1989 and 2001 he taught biostatistics and analytic epidemiology at the Department of Population Medicine, University of Guelph, Ontario, Canada where he held the position of full professor with tenure. At present, he is the principal scientist and the chairman of the Department of Biostatistics, Epidemiology, and Scientific Computing at King Faisal Specialist Hospital and Research Center, and adjunct professor of Biostatistics at the Department of Epidemiology and Biostatistics, Schulich School of Medicine and Dentistry, University of Western Ontario, London, Ontario, Canada. Professor Shoukri has papers published in *Statistics in Medicine, Biometrics, Statistical Methods in Medical Research, Biostatistics, Applied Statistics, the Journal of Statistical Planning and Inference*, and many other journals. He is a senior author of *Statistical Methods for Health Sciences* (first and second editions) and *Statistical Analysis of Correlated Data with SAS and R* (third edition), all published by Chapman & Hall/CRC Press. He is a fellow of the Royal Statistical Society of London and an elected member of the International Statistical Institute.

1

General Introduction

The main focus of this book is on the evaluation of raters' agreement and reliability. We demonstrate that both measures are related to the subject of error in measurements. Throughout the book we emphasize application in medical and epidemiological research. It is known that some of these measurements are carried out on a continuous scale (e.g., blood pressure levels, brain size, body mass index, tumor size, etc.) or on a categorical scale (e.g., presence of a condition, severity of illness, tumor type, etc.). Another type of variable is count (e.g., the number of polyps in the colon, the number of freckles on the face, the number of typographical errors per page in a manuscript, etc.).

There are many classifications of types of variables, as each type requires a specific statistical technique for analysis and presentations. It is widely acceptable that measurements of the relevant variables be classified into, nominal, categorical, ordinal, and continuous. The issue of measurement error for continuous variables would affect the reliability indices, while for categorical variables are said to be misclassification errors, which would affect the levels of agreements among raters. The message is that while reliability and agreement are conceptually different, the parameters for the assessment of raters' reliability and agreement differ as well, according to the scale of measurement.

There is a vast literature on the science of measurements and the reader is referred to an earlier paper by Stevens (1946) and a very interesting study by Hand (1996). In Stevens' work, the issue of understanding measurements was discussed in both physical and social sciences. While in physical sciences the problem was straightforward, in social sciences and particularly in psychology, assigning measurements was not clear. In medicine, and as was noted by Hand (1996), measurement procedures can be very complex: "a prime example being attempts to formulate quality of life scales." A good illustration may be found in Bishop et al. (1992).

This chapter is divided into two parts. In Part I, we review the literature which outlines the importance of the concepts of reliability and agreement in medical research and other fields of research. In Part II, we provide the interested readers with the mathematical tools that are needed for the methodological developments of both reliability and agreement. Those who are not interested may skip Part II and proceed to the remaining chapters.

1.1 Review

1.1.1 Measurement Errors in Medical Research

Researchers in many fields have become increasingly aware of the problem of errors in measurements. The investigations into the scientific bases of measurement errors began over one-and-a-half centuries ago by physical scientists and engineers. In clinical and medical research, measurement errors arise, in part, due to differences in diagnoses reported by clinicians, or differences in the accuracy of the measuring devices used by medical laboratories, or differences in the background training of medical technologists. Similar facts are widely known among scientists, and particularly clinical chemists who have spent a great deal of effort attempting to decipher the conflicting results related to blood constituents. There are numerous examples in clinical medicine that illustrate these situations. These are as follows.

In radiology, inconsistencies and inaccuracies have long been known to exist in the readings of chest radiographs. For example, Birkelo et al. (1947) found that in five readers who attempted to select radiographs suggesting pulmonary tuberculosis from a largely normal group, none succeeded in selecting more than 75% of the abnormal films. Moreover, when the same films were read again after 3 months, a reader was likely to change his mind once in every five of the positive cases. Fletcher and Oldham (1964) revealed that when they tested the ability of different observers to grade and classify cases of pneumoconiosis, a lack of consistency both between observers and in the same observer persisted. Another example of interobserver inconsistency was reported by Yerushalmy et al. (1950). They asked six experienced readers to state whether a good film was unchanged, better, or worse. All six readers agreed over only two-third of the pairs. Analysis showed that the disagreement was not confined to films of poor technical quality, and that unilateral disease was so tricky to classify as bilateral. On reading the films, a reader was likely to disagree with his previous interpretation of the radiographic appearances once in about every five cases, and once in 14 cases he would change his own reading of the second film from better to worse or vice versa.

Another example is given in an international pathology study that was conducted to measure the agreement among and within groups of pathologists involved in the categorization of childhood rhabdomyosarcoma according to four pathology classifications. Asmar et al. (1994) used data concerning agreement and survival experience according to patho-new subtypes as the basis for the selection of a proposed new pathological classification. A random sample comprising 800 eligible patients was selected. Pathologists representing eight institutions reviewed 20% of the patients. The objective was to determine the level of agreement with their original classification. In each

instance, the patients were classified according to four pathology systems:
the conventional system, the International Society of Pediatric Oncology system, the National Cancer Institute (NCI) system, and the cytohistological
system. The study reported low agreement among the pathologists for the
four systems. For reproducibility within the groups of pathologists, higher
levels of agreements were reported.

This issue of agreement among pathologists in histological diagnosis is of
paramount importance. That different pathologist, sometimes reach different conclusions when examining the same histological system is well known.
Sisson (1975) attempted to explore the factors responsible for differences of
observation and interpretation between pathologists in connection with histological diagnosis. The differences among pathologists were attributed to
poorly stated questions, unclear definition of agreement or disagreement,
and probably lack of control over the conditions under which the study is
conducted. Finally, "appropriate statistical methods" must be employed to
enable the correct conclusions to be drawn from the observations.

Apparently, radiologists are easy to test, but clinicians, even in a simple
examination of the chest, show the most remarkable variation in the frequency with which they elicit physical signs. This was made clear when eight
experienced clinicians—all Fellows of the Royal College of Physicians—
examined the chests of 20 patients with emphysema (Fletcher, 1952). Adhering
to the ordinary simple signs taught to students, they were never unanimous
in a single instance; and most signs produced only two-third agreement.
There is no doubt that such disagreement among clinicians does not pro-
mote confidence in physical examination. A similar lack of conformity has
been found in many instances. In the assessments of Mantoux tests, the general state of a child's nutrition, the redness of the throat, or the degree of
cyanosis or anemia have all revealed discrepancies. One should note that the
issue of disagreement goes beyond error in diagnosis. For example, Cochrane
et al. (1951) found that taking medical history is very liable to be biased by
the attitude of the recorder and that answers to even the simplest questions
are not always reproducible. This issue is illustrated as an example by
Cochrane et al. (1951).

In large-scale surveys of coalminers carried out by Cochrane et al. (1951),
who interviewed different groups of men, the age distributions of the groups
were similar. All the observers asked the miners in their groups the same
question. The answers to some of these questions showed that the observers'
bias influenced the frequency with which a positive answer was recorded.
Other questions were more consistently answered in the different groups. It
was suggested, after discussing the results, that, more attention should be
paid to possible lack of reproducibility in the answers that patients give to
questions commonly asked in clinical practice, and that, for research purposes, answers that are not reproducible are worthless.

The disagreement among clinicians in making medical diagnoses is a serious problem and is widespread. Disagreement may be due to the errors of

observation where certain abnormalities are wrongly identified or even missed altogether. The practical significance of observer variation has often been disputed and for a long period of time, its importance in radiology has been accepted. However, in an important contribution by Davies (1957), it was shown that an electrocardiogram is subject to observer variation, and is as difficult to interpret as is a chest film. They are of equal importance, in that one is as valuable in the diagnosis of heart disease as is the other in the diagnosis of pulmonary disease; and both are single investigations upon which diagnosis and treatment may largely depend. Davies' study included 100 tracings: half had been reported routinely to show infarction, a quarter to be normal, and a quarter to show various abnormalities other than infarction. Nine experienced readers reported their opinions of these electrocardiograms on two separate occasions. They were allowed the choice of one of three reports normal, abnormal, or infarction. Complete agreement was reached in only one-third of the 100 tracings, majority agreement in half, but there was considerable dispute about one tracing in five. After the second reading, it was found that on an average, the reader disagreed with one in eight of their original reports. This considerable observer variation affected the normal, abnormal, and infarction tracings equally; it was much larger than it had been expected and must represent the unrecognized difficulties of electrocardiographic diagnosis. Apparently, tracings from the intermediate zones are of little or no diagnostic value, but are very likely to be interpreted according to the clinical bias. In this way, observer variation may add to diagnostic error.

The clinical importance of this variation might be debatable, but as was noted by Davis, it is so large that in the absence of reliable information its importance cannot be denied.

Medical imaging is another area where the investigations of accuracy and interobserver variability were conducted. Jung et al. (2004) designed a study to investigate the accuracy and interobserver variability of a standardized evaluation system for endorectal three-dimensional (3D) magnetic resonance (MR) spectroscopic imaging of the prostate. Endorectal MR imaging and MR spectroscopic imaging were performed in 37 patients before they underwent radical prostatectomy. Two independent spectroscopists (observers or raters) scored the spectra of the selected voxels of the sampled patients, on a scale from 1 (benign) to 5 (malignant). The raters demonstrated an excellent interobserver agreement in this study. This excellent agreement demonstrated the potential effectiveness of metabolic information to identify prostate cancer, and the clinical usefulness of this system supports testing in clinical trials of MR imaging with MR spectroscopic imaging.

Applications of the concept of interrater agreement in breast cancer tumor grading have been considered by Fanshawe et al. (2008). Due to the fact that breast cancer is a heterogeneous disease, variability in shape, size, and character is quite high and that the morphological structure of the tumor is related to the degree of malignancy that is recognized by oncology pathologioto. In addition to the oxicting mothodc to accecc agroomont among multiple raters, Fanshawe et al. (2008) proposed several methods to reflect the extent to which the distribution of ratings provided by individual observers agrees with that provided by all observers. One method termed by them, "agreement score method," is compared with a Bayesian latent trait model. The "agreement score method" is a simple, nonchance corrected statistics that can be easily calculated and potentially used in order to provide some evidence whether there may be raters whose behavior is discrepant compared to that of the majority. Fanshawe et al. (2008) regarded this index as a measure of the relative agreement between an individual rater and a population of raters. This agreement score was interpreted by them as the probability that a given rater will agree with another randomly selected rater on a randomly selected sample. The Bayesian latent trait model has an advantage in case there is missing rating information and the number of raters is large. However, one has to impose the strong assumption that the missing information is completely at random (MCAR).

Scoring method techniques to assess the agreement between two or more raters, is ubiquitous in breast cancer tumor research. Another example in this area is the study by Diaz et al. (2004). Estrogen receptor (ER) status is useful clinically as a predictor, and positive ER status identified breast cancer patients who are most likely to benefit from hormonal therapy. The ER expression level of an individual tumor correlates with the degree and likelihood of response to hormonal therapy. To study the agreement among rates and with image analysis for ER immunohistochemical scoring, 70 patients and three pathologists were included. The authors demonstrated a high level of agreement between manual scoring with the ER determined by image analysis. Among the three pathologists, the interobserver agreement was almost perfect. The authors concluded that immunohistochemical evaluation of ER expression in breast cancer by image analysis is comparable to that by manual scoring and that interobserver variability among pathologists for manual scoring was "minimal."

Another example of interrater agreement is found in cardiology. A study was designed by Solm et al. (2004) with the aim to perform a head-to-head comparison between single-photon emission computer tomography (SPECT) and cardiovascular magnetic resonance (CMR), to evaluate hemodynamic significance of angiographic findings in bypass grafts. The rationale was, "the hemodynamic significance of a bypass graft stenosis may not always accurately be determined from the coronary angiogram." The study was conducted on 75 arterial and vein grafts from 25 patients. They were evaluated by SPECT and CMR, and the complete evaluation was obtained in 46 grafts. The agreement was good. The authors concluded that CMR may offer an alternative method to SPECT for the functional characterization of angiographic lesions (Table 1.1).

Another interesting example from the radiology literature is a study conducted by Dhingsa et al. (2002), whose objective was to assess the agreement

TABLE 1.1

Cross-Classification of CMR with SPECT for 46 Grafts

	CMR		
SPECT	**CFVR ≥ 2.0**	**CFVR < 2.0**	**Total**
Normal perfusion	18	2	20
Abnormal perfusion	7	19	26
Total	25	21	46

Note: CFVR = Coronary flow velocity reserve.

between general practitioners (GPs) and radiologists as to whether a radiation exposure of patients is justified.

This study was conducted to assess the agreement between three consultant radiologists and three GPs. The criteria for evaluation were based on information given on a request card and whether the request conforms to the Royal College of Radiologists' (RCR) guidelines. The consultant radiologists consisted of a musculoskeletal radiologist, a cross-sectional radiologist and a senior lecturer. The GP group consisted of a senior lecturer and two senior GPs. The reviewers were asked to state whether they felt that the request justified a radiation exposure, and whether the request conformed to the RCR guidelines. The results of the study showed that the greatest agreement between physicians is when guidelines are used and least agreement is on using only the request form.

The need for error-free diagnoses has been the subject of intense discussion among medical decision makers. This discussion has been fueled by both economical and ethical considerations. The countries where medical care is a publicly provided good suffer from an increase in the cost of health care, and a steady reduction in government spending in this area. It is then clear that unnecessary repeated and costly testing should be avoided in order to ensure efficient delivery of care. Likewise, under free enterprise health care systems, prudent fiscal policies of Health Maintenance Organization (HMO) mandate physicians to prescribe only one or two reliable tests, so that decision regarding treatment modality is correctly made, with the least cost.

In view of this, all laboratory tests must be validated before being introduced for patient testing to ensure that the reported measurements (e.g., blood glucose levels) will meet a desired degree of reliability. Validating a new technique begins with the consideration for, and selection of a new test method for the patient's use. Evaluating the analytic performance of medical tests is required to assess the degree of error expected due to inaccuracy and imprecision and to confirm that the level of errors is bounded primarily by clinical requirements.

In the next section, we introduce the concepts of reliability and agreement.

1.1.2 Reliability and Agreement

The terms "reliability" and "agreement" are often used interchangeably. However, they are conceptually different. Reliability of a measuring instrument lies in its ability to differentiate among subjects. It is the ratio of the subjects' variance to the total variance (Shrout and Fleiss, 1979). As for example, a clinician who provides the correct diagnosis all the time for all his patients will not be considered reliable unless he can distinguish the healthy from the diseased. On the other hand, agreement refers to conformity. The index of agreement determines whether the same value is achieved if a measurement is performed more than once by the same rater or by different raters.

In general, the concept of reliability is best explained by considering a very simple model for the outcome of interest as

$$Y = T + E.$$

The random variable Y represents a measurement, T is the true score representing the average value of all measurements within a patient, and E is the measurement error, assumed to be uncorrelated with T.

Under this assumption we have, $\text{var}(Y) = \text{var}(T) + \text{var}(E)$. The reliability of a measuring instrument is defined as

$$\rho = \frac{\text{var}(T)}{\text{var}(Y)} = \frac{\text{var}(T)}{\text{var}(T) + \text{var}(E)}. \qquad (1.1)$$

It is clear that when the patients are from a heterogeneous population (i.e., $\text{var}(T)$ is large) then ρ is large. The more heterogeneous the population, the larger is the reliability. On the other hand, when the population of patients is homogeneous (i.e., $\text{var}(T)$ is small), the reliability is lower. This means that ρ depends on the nature of the population from which we have sampled the patients. Note, also when $\text{var}(T)$ is fixed ρ decreases when the variance of error measurements $\text{var}(E)$ increases. That is, the error measurement has an attenuating effect on the reliability coefficient. In a typical reliability study, we assume that a random sample of k subjects is rated independently by n observers (raters). Shrout and Fleiss (1979) described three different designs from which one can measure reliability:

1. Each subject is rated by a different set of n raters, randomly selected from a large population of raters.
2. A sample of n rates is selected at random from the population of raters, and each rater measures each subject.
3. Each subject is measured by each of the same n raters. These raters are assumed to be the only available raters.

The above models are mathematically different. However, the common element among them is the blinded assessment of each subject by each rater.

That is, each rater should not know the result of assessments made by other raters. Moreover, in a situation where there is only one rater (test–retest reliability studies) that rater should be blinded regarding the result of his/her previous assessments. Moreover, each model attempts to decompose the total variation in the scores into possible effects, such as the rater's effect, the subject effect, and the possible interaction. Models for estimating reliability under the above situations are discussed in the next chapter. Under the above models, the intraclass correlation is used to measure rater reliability for continuous data.

Reliability models can be extended to generalizability theory. This theory is believed to have been originally introduced in an article by Cronbach et al. (1963). It differs from the well-known reliability theory where it attempts to decompose the measurement error component into several sources and isolate these major sources of errors so that a cost-efficient measurement design can be built (Shavelson et al., 1989). Therefore, rather than summing up all the sources of error in E, we should be able to know the amount of variation attributed to, say, gender, ethnicity, and time.

For nominal and ordinal categorical measurements, the kappa coefficients and its variants are used to measure the agreements among raters. Intraclass coefficients defined in Equation 1.1 may also be used for categorical data (Fleiss, 1981, 1986).

The debate among researchers on the difference between reliability and agreement took a philosophical slant. Guyatt et al. (1987) emphasized the distinction between reliability and agreement indices. They argued that reliability parameters are required for instruments that are used for discriminative purposes, while agreement parameters are used for evaluative purposes. Through a hypothetical example, they demonstrated that discriminative instruments should possess a high level of reliability. Thus, if var(T) is large, a certain level of measurement error is tolerable. For an evaluative measurement instrument between the subjects variation or the degree of homogeneity in the population of subjects is of no concern, rather the measurement error is important. These issues are elaborated in subsequent chapters.

There are inherent statistical issues that must be dealt with in our attempt to understand the sources of errors in clinical measurements. One needs to understand the types of errors (random or systematic) and how to measure their magnitude, and if the degree of error affects the interpretation and possibly patient's care. Clearly, if the potential error is large enough to lead to misdiagnosis, then the measuring instrument is not acceptable.

In the next chapters of this book, we focus on the types of experimental designs that are needed to obtain best estimates of reliability indices of measuring devices. In Chapter 2, we focus on interval scale measurements, introduce the appropriate indices of their reliability, and examine some of their statistical properties. Chapter 3 is devoted to the method comparison study. Chapter 4 investigates the coefficient of variation and the within-subject coefficient of variation as measures of reproducibility. Analysis of raters' agreement for categorical data is discussed in Chapter 5, and the model based approaches to the analysis of raters' agreement in multiple categories are discussed in detail in Chapter 6.

1.2 Probability and Its Application in Medical Research

1.2.1 Elements of Probability

In this section, the terms probability, relative frequency, proportion, and rate are used synonymously. If A denotes the event that a randomly selected individual from a population has defined characteristics (e.g., myocardial infarction), then $P(A)$ denotes the proportion of all subjects who have the characteristics. Classically, probability can be defined as follows: given a set of equally possible outcomes of an experiment the probability of an event, denoted by $P(A)$ is

$$P(A) = \frac{\text{Number of outcomes comprising } A}{\text{Total number of possible outcomes}}. \qquad (1.2)$$

This definition assumes that we can perform infinitely many replications of the same experiment.

1.2.2 Some Rules of Probability

Rule 1: Addition Rule

If A and B are two events, a question may be: What is the probability of either A or B occurring? Here is a simple example. Suppose a group of 200 subjects can be classified according to their ethnicity such that, 20 are Black, 50 are White, 40 Hispanic, and the rest are Asian. If one subject's name is drawn randomly, what is the probability of being either White or Asian? The answer is 140/200. Notice that the two events, "White" and "Asian," are mutually exclusive, in that when we draw a name from the list of names, it cannot be both White and Asian, and if A and B are mutually exclusive event, then

$$P(A \text{ or } B) = P(A) + P(B). \qquad (1.3)$$

Hence,

$$P(\text{White or Asian}) = P(\text{White}) + P(\text{Asian})$$

$$= \frac{50}{200} + \frac{90}{200} = \frac{140}{200}.$$

Now consider the extension of the above example whereby the Black subjects are such that 10 are men, 10 are women; for the White subjects, 30 are men and 20 are women; Hispanic subjects are 20 men and 20 women; and Asian subjects are 50 men and 40 women. If one individual is drawn from the list at random we would like to know the probability of being either Black or a woman. In this case, the event ethnicity and gender are not mutually exclusive, because the individual could be both Black and a woman; therefore, we cannot use the addition rule (Equation 1.3). How many subjects are either Black or women? There are 20 Black + 90 women—110 subjects, one notices that Black women have been counted twice—once as Black and once as a woman; and therefore, Black women must be subtracted to get the correct count. Therefore,

$$P(\text{Black or woman}) = P(\text{Black}) + P(\text{woman}) - P(\text{Black and woman})$$

$$= \frac{20}{200} + \frac{90}{200} - \frac{10}{100} = \frac{1}{2}.$$

The rule for any two events A and B is

$$P(A \text{ or } B) = P(A) + P(B) - P(A \text{ and } B). \tag{1.4}$$

In the special case in which A and B are mutually exclusive event,

$$P(A \text{ and } B) = 0. \tag{1.5}$$

Rule 2: Multiplication Rule

Conditional Probability

We have just seen the probability that the events A and B jointly occur is written as $P(A \text{ and } B)$ or $P(AB)$, or mathematically expressed as $P(A \cap B)$. This is known as the joint probability of A and B. To evaluate the probability $P(AB)$, we use the notion of conditional probability: the probability of an event occurring; given that another event has already occurred. The notation is:

$P(A|B)$ = Probability that a randomly selected individual has characteristic A, given that he/she has characteristic B or conditional on his/her having characteristic B.

Let $P(B)$ represent the proportion of all people who possess characteristic B, and let $P(A \text{ and } B)$ represent the proportion of all people who possess both the characteristics A and B. Then, by definition, provided that $P(B) \neq 0$,

$$P(A|B) = \frac{P(A \text{ and } B)}{P(B)}. \tag{1.6}$$

Similarly, provided that $P(A) \neq 0$,

$$P(B|A) = \frac{P(A \text{ and } B)}{P(A)}. \tag{1.7}$$

By the association of two characteristics, we mean that when a person has one of the characteristics, say B, his chances of having the other are affected. By the independence or lack of association of two characteristics we mean that if a person has one of the characteristics, it does not affect his chances of having the other. Thus, if A and B are independent, then the rate at which A is present specific to people who possess B, $P(A|B)$, is equal to the overall rate at which A is present, $P(A)$. By Equation 1.6, this implies that

$$\frac{P(A \text{ and } B)}{P(B)} = P(A), \tag{1.8}$$

or

$$P(A \text{ and } B) = P(A)P(B). \tag{1.9}$$

This is the definition of two independent events.

By $P(A|B)$ we mean the proportion out of all people who have characteristic B who also have characteristic A, so that both the numerator and the denominator of $P(A|B)$ must be specific to B.

The so-called Bayes' theorem gives the inverse of the above probability:

$$P(B|A) = \frac{P(A|B)P(B)}{P(A)}$$

1.2.3 Discrete Random Variables

1.2.3.1 Univariate Functions

A discrete random variable Y is a function from the sample space Ω that has a finite or countable infinite number of real numerical values. For the random variable Y, each value y has a probability $\Pr[Y = y]$, known as the probability distribution of Y. The expected value of Y is

$$\mu = E(Y) = \sum_y y \Pr[Y = y].$$

If $g(Y)$ is a real-valued function of Y, then

$$E[g(Y)] = \sum_y g(y)\Pr[Y = y].$$

The variance of Y is

$$\sigma_y^2 = \text{var}(Y) = \sum_y (y - \mu)^2 \Pr[Y = y]$$

$$= E(Y^2) - E^2(Y).$$

1.2.3.2 Bivariate Functions

Let X, Y be two discrete random variables. The joint probability function is $\Pr[X = x, Y = y]$.

The marginal probability distribution of X is

$$\Pr[X = x] = \sum_y \Pr[X = x, Y = y]$$

and of Y is

$$\Pr[Y = y] = \sum_x \Pr[X = x, Y = y].$$

Expectations:

1. $E(X \pm Y) = E(X) \pm E(Y) = \mu_x \pm \mu_y$.
2. $\text{cov}(X, Y) = E[(X - \mu_x)(Y - \mu_y)]$ is called the covariance of the random variables.
3. The correlation between X and Y is

$$\rho = \text{Corr}(X, Y) = \frac{\text{cov}(X, Y)}{\sqrt{\text{var}(X)\text{var}(Y)}}. \tag{1.10}$$

4. $\text{var}(X \pm Y) = \text{var}(X) + \text{var}(Y)$, provided that X and Y are independent. When X and Y are correlated, then

$$\text{var}(X \pm Y) = \text{var}(X) + \text{var}(Y) \pm 2\rho\sqrt{\text{var}(X)\text{var}(Y)}. \tag{1.11}$$

If X and Y are independent, then $\rho = 0$, the converse is not true.
5. If X and Y are independent, then

$$E(YX) = E(X) \cdot E(Y).$$

1.2.3.3 Bernoulli Trials

An experiment of a particularly simple type is one in which there are only two possible outcomes, such as disease or health, exposed or nonexposed. For mathematical convenience we designate the two possible outcomes of such an experiment as 1 and 0.

Definition: 1.1

A random variable Y has a Bernoulli distribution with parameter $0 \le P \le 1$ if Y can take the values 1 and 0, with probabilities $P(Y = 1|p) = p$, and $P(Y = 0|p) = q = 1 - p$.

The probability function of Y can be written as

$$P(Y = y|p) = \begin{cases} p^y q^{1-y} & \text{for } y = 0,1 \\ 0 & \text{elsewhere} \end{cases}.$$

One can show that

$$E(Y|p) = p \quad \text{and} \quad \text{var}(Y|P) = p(1 - p).$$

1.2.3.4 Binomial Distribution

If the random variables Y_1, Y_2, ..., Y_n are independently and identically distributed outcomes of n Bernoulli trials with parameter p and if $Y = Y_1 + Y_2 + \cdots + Y_n$, then Y has a binomial distribution with parameters n and p with probability function

$$P(Y = y|p) = \begin{cases} \binom{n}{y} p^y q^{n-y} & y = 0, 1, 2, ..., n \\ 0 & \text{elsewhere} \end{cases}.$$

One can show that

$$E(Y|p) = E(Y_1 + Y_2 + \cdots + Y_n) = np$$

$$\text{var}(Y|p) = \text{var}(Y_1 + Y_2 + \cdots + Y_n) = npq.$$

1.2.3.5 Multinomial Distribution

Suppose that a population can be partitioned into $k \ge 2$ mutually exclusive categories and that the proportion of objects or subjects in the population that belong to category i is P_i ($i = 1, 2, ..., k$). It is assumed that $P_i > 0$ for

$i = 1, 2, \ldots, k$ and that $P_1 + P_2 + \cdots + P_k = 1$. Furthermore, suppose that n subjects are selected at random from the population, with replacement; and let Y_i denote the number of subjects selected that belong to category $i (i = 1, 2, \ldots, k)$. Then, it is said that the random vector $Y = (Y_1, Y_2, \ldots, Y_n)$ has a multinomial distribution with parameters $(n, p_1, \ldots, p_k)$. The probability function of Y is

$$P(Y_1 = y_1, Y_2 = y_2, \ldots, Y_k = y_k) = \frac{n!}{y_1! y_2! \ldots y_k!} p_1^{y_1}, p_2^{y_2}, \ldots, p_k^{y_k}. \qquad (1.12)$$

One can show that

$$E(Y_i | P_i) = np_i, \quad \mathrm{var}(Y_i | P_i) = np_i q_i,$$

and

$$\mathrm{cov}(Y_i, Y_j | p_i, p_j) = -np_i p_j \quad (i \neq j = 1, 2, \ldots, k).$$

Note, for $k = 2$, the multinomial becomes a binomial distribution with parameters (n, p_i).

1.2.4 Continuous Random Variables

A continuous random variable Y with probability density function (PDF) f is a random variable such that for each interval A of real numbers

$$\Pr[Y \in A] = \int_A f(y) \, dy.$$

The expectation of Y is

$$\mu_y = E(Y) = \int_{-\alpha}^{\alpha} y f(y) \, dy.$$

The expectation of a function $g(Y)$ of a continuous random variable Y is

$$E[g(Y)] = \int_{-\alpha}^{\alpha} g(y) f(y) \, dy.$$

The variance of Y is

$$\sigma_y^2 = \text{var}(Y) = \int_{-\alpha}^{\alpha} (y - \mu_y)^2 f(y) \, dy$$

$$= E(Y^2) - E^2(Y).$$

1.2.4.1 Joint Densities

Consider continuous random variables X and Y with densities $f(x)$ and $f(y)$. If their joint density $f(x,y)$ is defined for every rectangle R, the probability is

$$\Pr[(X,Y) \in R] = \iint_R f(x,y) \, dx \, dy.$$

The marginal PDF of X, denoted by $f(x)$ is

$$f(x) = \int_{-\infty}^{\infty} f(x,y) \, dy.$$

Similarly,

$$f(y) = \int_{-\infty}^{\infty} f(x,y) \, dx.$$

If $z = h(x,y)$ is a real-valued function of X and Y, then

$$E(z) = \int_{-\infty}^{\infty} \int_{-\infty}^{\infty} zf(x,y) \, dx \, dy.$$

Similar to the discrete random variable, we have

$$\text{var}(X \pm Y) = \sigma_x^2 + \sigma_y^2 \pm 2\rho\sigma_x\sigma_y.$$

where ρ is the correlation coefficient between X and Y and is generally defined by

$$\rho = \frac{\text{cov}(x,y)}{\sigma_x\sigma_y}.$$

If X and Y are independent, then they are uncorrelated ($\rho = 0$). Hence,

$$\text{var}(X \pm Y) = \text{var}(X) + \text{var}(Y).$$

Finally, if X and Y are independent, and if $g(X)$ and $h(Y)$ are real-valued functions, then

$$E[g(X) \cdot h(Y)] = E[g(X)] \cdot E[h(Y)].$$

1.2.4.2 β-Distribution

A continuous random variable x is said to have a β-distribution with parameters $\alpha > 0$ and $\beta > 0$ of its PDF is given by

$$f(x|\alpha,\beta) = \begin{cases} \dfrac{\Gamma(\alpha + \beta)}{\Gamma(\alpha)\Gamma(\beta)} x^{\alpha-1}(1 - x)^{\beta-1} & 0 < x < 1 \\ 0 & \text{elsewhere} \end{cases}. \tag{1.13}$$

It can be shown that

$$E(X^r) = \frac{\Gamma(\alpha + \beta)\Gamma(\alpha + r)}{\Gamma(\alpha)\Gamma(\alpha + \beta + r)}.$$

Therefore,

$$E(X) = \frac{\alpha}{\alpha + \beta}, \tag{1.14}$$

and

$$\text{var}(X) = \frac{\alpha\beta}{(\alpha + \beta)^2(1 + \alpha + \beta)}. \tag{1.15}$$

1.2.4.3 β-Binomial Distribution

Note that we have written the probability function of the binomial distribution conditional on a given value of the probability p. Assume that p has the PDF of a β-distribution; that is

$$f(p|\alpha,\beta) = \frac{\Gamma(\alpha + \beta)}{\Gamma(\alpha)\Gamma(\beta)} p^{\alpha-1}(1 - p)^{\beta-1}.$$

The unconditional distribution of the binomially distributed random variable Y is obtained by average over the distribution of the then random variable p. Hence,

$$P(Y = y) = \int_0^1 P(Y = y|p)f(p|\alpha,\beta)\,dp$$

$$= \binom{n}{y}\frac{\Gamma(\alpha + y)\Gamma(n + \beta - y)\Gamma(\alpha + \beta)}{\Gamma(n + \alpha + \beta)\Gamma(\alpha)\Gamma(\beta)}. \quad (1.16)$$

This probability function is known as the β-binomial distribution (BBD). The mean of the BBD is

$$E(Y) = E[np] = \frac{n\alpha}{\alpha + \beta}.$$

The variance of the BBD is

$$var(Y) = E\left[var(Y|p)\right] + var\left[E(Y|p)\right]$$

$$= E[nP(1 - p)] + var[np]$$

$$= \frac{n\alpha\beta(n + \alpha + \beta)}{(\alpha + \beta)(1 + \alpha + \beta)}. \quad (1.17)$$

If we write $\rho = (1 + \alpha + \beta)^{-1}$, $\pi = \alpha(\alpha + \beta)$, then Equation 1.17 may be written as

$$var(Y) = n\pi(1 - \pi)[1 + (n - 1)\rho].$$

1.2.4.4 Dirichlet Distribution

The Dirichlet distribution (DD) may be considered as a multivariate generalization of the β-distribution. Let $P = (P_1, P_2, ..., P_k)$ denote a random vector whose elements sum to 1. Conditional on a parameter vector $\alpha = (\alpha_1, \alpha_2, ..., \alpha_k)^t$, the Probability density function (Pdf) of P is

$$f\left(\underline{P}|\underline{\alpha}\right) = \frac{\Gamma\left(\sum_k \alpha_k\right)}{\prod_k \Gamma(\alpha_k)} \prod_k P_k^{\alpha_k - 1},$$

where $P_k > 0$, and $P_1 + P_2 + \cdots + P_k = 1$. Let $\alpha_0 + \alpha_2 + \cdots + \alpha_k$; then

$$E(P_i) = \frac{\alpha_i}{\alpha_0}$$

$$\text{var}(P_i) = \frac{\alpha_i(\alpha_0 - \alpha_i)}{\alpha_0^2(\alpha_0 + 1)}$$

$$\text{cov}(P_i, P_j) = \frac{-\alpha_i\alpha_j}{\alpha_0^2(\alpha_0 + 1)}.$$

The marginal distribution of P_i is the univariate β-distribution with parameters $(\alpha_i, \alpha_0 - \alpha_i)$.

1.2.4.5 Multinomial Dirichlet Distribution

The multinomial Dirichlet distribution (MDD) is obtained in a similar manner by which we obtained the BBD. First, we assume that conditional on the random vector $\underline{P}$, the vector $\underline{Y} = (y_1, ..., y_k)$ has a multinomial distribution. Second, we assume that P has a DD. Therefore, the unconditional distribution of $\underline{Y} = (y_1, ..., y_k)$ is

$$P(y_1,...,y_k) = \int P(Y|P) f(P|\alpha) \, dP$$

$$= \frac{n!\,\Gamma(\alpha_0)}{\prod\limits_{i=1}^{k} y_i!\,\Gamma\left(\sum\limits_{i=1}^{k}(y_i + \alpha_i)\right)} \prod_{i=1}^{k}\left\{\frac{\Gamma(y_i + \alpha_i)}{\Gamma(\alpha_i)}\right\}. \tag{1.18}$$

The joint distribution $P(y_i, ..., y_k)$ defines the so-called MDD.

1.2.5 Multivariate Normal Theory and Techniques of Inference

1.2.5.1 Normal Distribution

The normal distribution plays a fundamental role in the theory of statistical inference due, in part, to the importance of the central limit theory. The random variable X is defined to have a univariate normal distribution with a mean μ and variance σ^2 if and only if the PDF of X is given by

$$f(x, \mu, \sigma^2) = \frac{1}{\sigma\sqrt{2\pi}} \exp\left[\frac{\left(-(x - \mu)^2\right)}{2\sigma^2}\right]. \tag{1.19}$$

If $X_1, X_2, \ldots, X_n$ are independently and identically distributed normal random variables then $Q = \sum_{i=1}^{n}((X_i - \mu/\sigma))^2$ has a chi-square distribution with n degrees of freedom.

If the p-dimensional random vector $\underline{X} = (X_1, \ldots, X_p)$ is distributed as multivariate normal random variable, then its PDF is given by

$$f\left(\underline{x}; \underline{\mu}, \Sigma\right) = \frac{(2\pi)^{-P/2}}{\Sigma 1^{-1/2}} \exp\left[-\frac{1}{2}(x-\mu)' \Sigma^{-1}(x-\mu)\right]. \tag{1.20}$$

We write $\underline{x} \sim N(x; \mu, \Sigma)$. It is assumed that Σ has rank p. The following results are given in Cramer (1946):

- The quadratic form $Q = (x - \mu)' \Sigma^{-1}(x - \mu)$ has a chi-square distribution with p degrees of freedom.
- The matrix Σ is the variance–covariance matrix of $\underline{X} = (X_1, \ldots, X_p)$, whose ith diagonal element is $\text{var}(X_i) = \sigma_i^2$ and off diagonal elements $\sigma_{ij} = \text{cov}(X_i, X_j)$.
- Let the P dimensional random vector X be distributed as $N(x; \mu, \Sigma)$. Partition X, μ, and Σ as

$$X = \begin{pmatrix} x_1 \\ x_2 \end{pmatrix}, \quad \mu = \begin{pmatrix} \mu_1 \\ \mu_2 \end{pmatrix}, \quad \Sigma = \begin{pmatrix} \Sigma_{11} \Sigma_{12} \\ \Sigma_{21} \Sigma_{22} \end{pmatrix}$$

where x_1 and μ_1 are p_1-column vectors, Σ_{11} is a square matrix of order p_1, and the size of the remaining vectors and matrices are thus determined. The random vector x_1 is normally distributed with mean μ_1 covariance matrix Σ_{11}, that is $x_1 \sim N(x_1; \mu_1, \Sigma_{11})$.
- The conditional distribution of x_1 given $x_2 = c_2$, where c_2 is a vector of constants, is normal with mean $\mu_1 + \Sigma_{12}\Sigma_{22}^{-1}(x_2 - \mu_2)$ and covariance matrix $\Sigma_{11.2}$, were $\Sigma_{11.2} = \Sigma_{11}\Sigma_{22}^{-1}\Sigma_{21}$.

Remark

When $p = 2$, we get the PDF of the bivariate normal distribution

$$f\left(x_1, x_2\right) = \frac{1}{2\pi\sigma_1\sigma_2\sqrt{1-\rho^2}} \exp\left[\frac{z_1^2}{\sigma_1^2} + \frac{z_2^2}{\sigma_1^2} - 2\rho\frac{z_1 z_2}{\sigma_1\sigma_2}\right] \tag{1.21}$$

where $z_i = x_i - \mu_i$; $i = 1, 2$, and $\rho = \text{Corr}(x_1, x_2)$.

1.2.5.2 Approximate Methods

1.2.5.2.1 Univariate Case

It is known from the elementary statistical theory that given the exact moments of random variables, one can find the exact moments of any linear combinations of these random variables. For example, if X_1 and X_2, are two random variables, then $Y = a_1 X_1 + a_2 X_2$ has mean

$$E(Y) = a_1 E(X_1) + a_2 E(X_2)$$

and variance

$$\text{var}(Y) = a_1^2 \text{ var}(X_1) + a_2^2 \text{ var}(X_2) + 2a_1 a_2 \text{ cov}(X_1, X_2).$$

In many practical situations, we may need a variance expression for non-linear function of X_1 and X_2. For example, we need to evaluate $\text{var}(X_1/X_2)$, or $\text{var}\left((X_1 - X_2)/\left(\sqrt{X_1^2 + X_2^2}\right)\right)$. When the exact variance cannot be obtained in a closed form, we resort to one of the most commonly used techniques known as the delta method (DM) (see Stuart and Ord, 1987). Roughly, for some sequence of random variables x_n with variance σ_n^2, with $\sigma_n \to 0$, then for a real-valued function g, differentiable at μ, with $(\partial\, g(\mu)/\partial\mu) \neq 0$, when $\mu = E(X_n)$, then to the first order of approximation

$$\text{var}\left(g\left(x\right)\right) = \left[\frac{\partial g(\mu)}{\partial\mu}\right]^2 \sigma_n^2. \tag{1.22}$$

For example, if X has a binomial distribution with $E(x/n) = P$, and $\text{var}(x/n) = (P(1 - P)/n)$, then $\text{var}(\log(x/n)) = (1 - P/nP)$.

1.2.5.2.2 Multivariate Case

Let $h(X_1, X_2, \ldots, X_n)$ be a real-valued function of $(X_1, X_2, \ldots, X_n)$, differentiable at $(\mu_1, \ldots, \mu_n)$ with $(\partial h/\partial x_i) \neq 0$ $(i = 1,2, \ldots, n)$, where $\mu_i = E(X_i)$; then

$$\text{var}(h) = \sum_{i=1}^{n} \text{var}(x_i)\left(\frac{\partial h}{\partial x_i}\right)^2 + \sum_{i \neq j}\Sigma \text{cov}(x_i, x_j)\left(\frac{\partial h}{\partial x_i}\right)\left(\frac{\partial h}{\partial x_j}\right) \tag{1.23}$$

and the partial derivatives are evaluated at $x_i \equiv \mu_i$.

Generally, if $h_r(X_1, X_2, \ldots, X_n)$, then under similar conditions and to the first order of approximation, we have the covariance between the functions of random variables

$$\text{cov}(h_r, h_s) = \sum_{i=1}^{n} \left(\frac{\partial h_r}{\partial x_i}\right)\left(\frac{\partial h_s}{\partial x_j}\right)\text{var}(x_i) + \sum\sum_{i \neq j}\left(\frac{\partial h_r}{\partial x_i}\right)'\left(\frac{\partial h_s}{\partial x_j}\right)'\text{cov}(x_i, x_j). \quad (1.24)$$

As an example, let $X = (X_1, X_2, X_3)$ be multinomially distributed random vector, with $E(X_i) = nP_i$, $\text{var}(X_i) = nP_i(1 - P)$, and $\text{cov}(X_i, X_j) = -nP_iP_j$.
Let $\theta_i = x_i/n$, $Y_1 = (\theta_1/\theta_1 + \theta_2)$, $Y_2 = (\theta_2/\theta_1 + \theta_2)$.
It can be shown by DM that

$$\text{var}(Y_j) = \frac{P_1P_2}{n(P_1 + P_2)^2}; \quad j = 1, 2$$

$$\text{cov}(Y_1, Y_2) = \frac{-P_1P_2}{n(P_1 + P_2)^3}.$$

1.2.6 Parametric Estimation of Population Parameters

1.2.6.1 Method of Moments

Let $Y_1, Y_2, \ldots, Y_n$ be a simple random sample of size n taken from a population whose probability distribution is characterized by $p \leq n$ parameters. We denote these parameters by $(\theta_1, \theta_2, \ldots, \theta_p)$. The simplest method of estimation of these parameters is the method of moments. The first step in using the method of moments in estimation is to construct as many sample moments as we have parameters:

$$m_r' = \frac{1}{n}\sum_{i=1}^{n} y_i^r \quad r = 1, 2, \ldots, p$$

The second step is to find the moments of the distribution from which the sample $\mu_r' = E(Y^r)$ is drawn. The population moments are assumed to be functions of $(\theta_1, \theta_2, \ldots, \theta_p)$, and therefore, we write them as $\mu_r'(\theta_1, \ldots, \theta_p)$. In the third step, we equate the sample moments to their respective population moments and solve the resulting equations for the unknown parameters:

$$\hat{\theta}_r = f_r(m_1', m_2', \ldots, m_p') \quad r = 1, 2, \ldots, p$$

where $f_r(m_1', m_2', \ldots, m_p')$ are functions of the sample moments.
As an example, let $y_1, y_2, \ldots, y_n$ be a simple random sample taken from the two parameters' normal population

$$f(y|\theta, \phi) = \frac{1}{\phi\sqrt{2\pi}}\exp\left[-\frac{1}{2\phi^2}(y - \theta)^2\right].$$

Since

$$E(Y) = \theta$$

$$E(Y^2) = \phi^2 + \theta^2,$$

we equate the above two equations to the sample moments $m_1' = \bar{y}$, $m_2' = 1/n \, \Sigma_{i=1}^{n} y_i^2$:

$$\hat{\theta} = \bar{y}.$$

$\hat{\phi}^2 + \hat{\theta}^2 = m_2'$, from which

$$\hat{\phi}^2 = m_2' - \bar{y}^2.$$

1.2.6.2 Moments of Sample Moments

Kendal and Ord (1989) gave very important results regarding the moments of the sample moment m_r'.

They showed that the sample rth noncentral moment is unbiased estimator of the population rth noncentral moment μ_r'. That is

$$E(m_r') = \mu_r'.$$

Moreover,

$$\text{var}(m_r') = \frac{1}{n}\left(\mu_{2r}' - \mu_r'^2\right) \tag{1.25}$$

and

$$\text{cov}\left(m_q', m_r'\right) = \frac{1}{n}\left(\mu_{q+r}' - \mu_q'\mu_r'\right). \tag{1.26}$$

Another useful result regarding the rth sample central moment m_r, where

$$m_r = \frac{1}{n}\sum_{i=1}^{n}\left(y_i - m_1'\right)^r$$

is given to the first order of approximation as

$$E(m_r) \doteq \mu_r = E(y - \mu_1')^r$$

and

$$\text{var}(m_r) \doteq \frac{1}{n}\left(\mu_{2r} - \mu_r^2 + r^2\mu_2\mu_{r-1}^2 - 2r\mu_{r-1}\mu_{r+1}\right), \tag{1.27}$$

$$\text{cov}\left(m_r, m_q\right) \doteq \frac{1}{n}\left(\mu_{r+q} - \mu_r\mu_q + rq\mu_2\mu_{r-1}\mu_{q-1} - r\mu_{r-1}\mu_{q+1} - q\mu_{r+1}\mu_{q-1}\right). \tag{1.28}$$

The above results were extended by Kendall and Ord to the bivariate case. For example, if there is a random sample $(X_1, Y_1), (X_2, Y_2), \ldots, (X_n, Y_n)$ from a bivariate population, then the bivariate sample moments are defined as

$$m_{rs}' = \frac{1}{n}\sum_{i=1}^{n} x_i^r y_i^s,$$

$$E\left(m_{rs}'\right) = \mu_{r,s}' = E[X^r Y^s],$$

$$\text{var}(m_{rs}') = \frac{1}{n}\left(\mu_{2r,2s}' - \mu_{r,s}'^2\right), \tag{1.29}$$

$$\text{cov}\left(m_{r,s}', m_{u,v}'\right) = \frac{1}{n}\left(\mu_{r+u,s+v}' - \mu_{r,s}'\mu_{u,r}'\right). \tag{1.30}$$

The results of this section will be utilized in subsequent chapters when properties of estimated measures of agreement and reliability are established.

1.2.6.3 Method of Maximum Likelihood

Based on the random sample, the joint PDF of $y_1, y_2, \ldots, y_n$ is formulated as

$$L = \prod_{i=1}^{n} f\left(y_1, y_2, \ldots, y_n \mid \theta_1, \theta_2, \ldots, \theta_p\right).$$

This function is called the likelihood function. It is maximized by setting the partial derivatives of the natural logarithm of this function, with respect to $(\theta_1, \theta_2, \ldots, \theta_p)$ zero, and then solve the resulting equations.

As an example, based on the random sample from the normal population, the likelihood function is

$$L = \left(\frac{1}{2\pi\phi^2}\right)^{n/2} \exp\left[-\frac{1}{2\phi^2}\sum_{i=1}^{n}(y_i - \theta)^2\right].$$

The natural logarithm of this function is

$$\ln L = -\frac{n}{2}\ln\left(2\pi\phi^2\right) - \frac{1}{2\phi^2}\sum_{i=1}^{n}(y_i - \theta)^2.$$

Differentiating the log-likelihood function with respect to the population parameters, we get

$$\frac{\partial \ln L}{\partial \theta} = \sum_{i=1}^{n}\frac{(y_1 - \theta)}{\phi^2}$$

$$\frac{\partial \ln L}{\partial \phi^2} = -\frac{n}{2\phi^2} + \frac{\sum_{i=1}^{n}(y_i - \theta)^2}{2\phi^4}.$$

Equating the above equations to zero and solving for θ and ϕ^2, we get

$$\hat{\theta} = \sum_{i=1}^{n}\frac{y_i}{n}$$

$$\hat{\phi}^2 = \sum_{i=1}^{n}\frac{\left(y_i - \bar{y}\right)^2}{n}.$$

1.2.7 Asymptotic Theory of Likelihood Inference

Case 1: Simple Null Hypothesis

Suppose that we would like to test the null hypothesis $H_0 : \theta = \theta_0$ against the alternative $H_a : \theta \neq \theta_0$, where θ_0 is not on the boundary of the parameter space of the vector $\theta = (\theta_1, \theta_2, ..., \theta_p)^T$. The likelihood ratio test was proposed by Neyman and Pearson (1928), and was based on the statistic

$$\lambda = 2l\left(\hat{\theta}; y\right) - 2l(\theta_0; y). \tag{1.31}$$

Here, $l(\theta; y)$ is the natural logarithm of the likelihood function, and $\hat{\theta}$ is the maximum likelihood estimator of θ. Define:

a. The score statistic $U(\theta) = (U_1(\theta), ..., U_p(\theta))^T$, where $U_i(\theta) = \partial l(\theta; y)/\partial\theta$
   $(i = 1, 2, ..., p)$.

b. The Fisher's information matrix $\Sigma(\theta)$, in which

$$\Sigma_{ij} = -E\left[\frac{\partial^2 l(\theta; y)}{\partial \theta_i \partial \theta_j}\right].$$

To test H_0 against H_a, Wald (1943) suggested the test statistic

$$Z^2 = \left(\hat{\theta} - \theta_0\right)' - \sum(\theta)\left(\hat{\theta} - \theta_0\right) \tag{1.32}$$

and Rao suggested

$$T^2 = \left\{U(\theta_0)\right\}' \sum{}^{-1}(\theta)\left\{U(\theta_0)\right\}. \tag{1.33}$$

Under the null hypothesis, the three test statistics (λ, z^2, T^2) are asymptotically distributed as chi-square with p degree of freedom.

As an example, let $Y_1, Y_2, ..., Y_n$ be a random sample taken from the normal distribution with mean m and variance ϕ^2. Suppose we would like to simultaneously test $H_0: m = m_0 \cap \phi^2 = \phi_0^2$ against H_1 : not H_0.

It will be left as an exercise to show that

$$\lambda = 2n \ln\left(\phi_0/\tilde{\phi}\right) + n\left(\overline{\phi}^2/\phi_0^2 - 1\right) \tag{1.34}$$

where

$$\tilde{\phi} = \left\{\Sigma(y_i - \overline{y})^2/n\right\}^{1/2},$$

$$\overline{\phi} = \left\{\Sigma(y_i - m_0)^2/n\right\}^{1/2},$$

$$z^2 = n\left(\overline{y} - m_0\right)^2/\tilde{\phi}^2 + 2n\left(\tilde{\phi} - \phi_0\right)^2/\tilde{\phi}^2,$$

$$T^2 = n\left(\overline{y} - m_0\right)^2/\phi_0^2 + n\left(\overline{\phi}^2 - \phi_0^2\right)/2\phi_0^4.$$

According to large sample theory of likelihood inference each of (λ, z^2, T^2) has chi-square with two degrees of freedom.

Case 2: Composite Null Hypothesis

Suppose that the parameter vector θ is partitioned so that $\theta = (\gamma', \beta')'$. It is assumed that θ is $k \times 1$ vector and β is a $q \times 1$ vector of nuisance parameters.

Let $H =: \gamma = \gamma_0$ against $H_1 : \gamma \neq \gamma_0$ be tested, while β remains unspecified. From Cox and Hinkley (1974), the likelihood ratio test is

$$\tilde{\lambda} = 2l\left(\hat{\theta}; y\right) - 2l\left(\gamma_0^T, \beta_0^T\right) \tag{1.35}$$

where $\hat{\theta} = \left(\hat{\gamma}^T, \hat{\beta}^T\right)^T$ is the maximum likelihood estimator of θ under the full model. Also, $\left(\bar{\gamma}_0, \hat{\beta}_0^T\right)$ is the maximum likelihood estimator of θ under the null hypothesis. The score statistic is obtained as follows:

Define

$$U_i(\theta) = \frac{\partial l(\theta)}{\partial \theta_i},$$

and the Fisher's information matrix as

$$I(\theta) = -E\left[\frac{\partial_2 l(\theta; y)}{\partial \theta_i \partial \theta_j}\right].$$

Now partition $Ut(\theta)$ and $I(\theta)$ so that

$$U = U(\theta) = \begin{pmatrix} U_\gamma \\ U_\beta \end{pmatrix}$$

$$I(\theta) = \begin{bmatrix} I_{11} & I_{12} \\ I_{21} & I_{22} \end{bmatrix}$$

where

$$I_{11} = -E\left[\frac{\partial^2 l}{\partial \gamma \partial \gamma^T}\right],$$

$$I_{12} = -E\left[\frac{\partial^2 l}{\partial \gamma \partial \beta}\right] = I_{21}^T,$$

$$I_{22} = -E\left[\frac{\partial^2 l}{\partial \beta \partial \gamma^T}\right].$$

Define

$$\Sigma(\theta) = I_{11} - I_{12}I_{22}^{-1}I_{21}.$$

Then, from Cox and Hinkley (1974), the Wald statistic is

$$\tilde{z}^2 = (\hat{\gamma} - \gamma_0)^T \sum (\theta)(\hat{\gamma} - \gamma_0) \tag{1.36}$$

and the score statistic is

$$\tilde{T}^2 = \{U(\hat{\gamma}_0)\}^T \left\{ \sum (\hat{\theta}) \right\}^{-1} \{U(\gamma_0)\}. \tag{1.37}$$

All the three statistics have chi-square distributions with q degrees of freedom.

1.2.8 Applications of Probability in Clinical Epidemiology

1.2.8.1 *Evaluation of Medical Screening Tests*

Medical screening programs are frequently used to provide estimates of prevalence where prevalence is defined as the number of diseased individuals at a specified point in time divided by the number of individuals exposed to risk at that point in time. Estimates of disease prevalence are important in assessing disease impact, in the delivery of health care. Since screening tests (STs) are less than perfect, estimated prevalence must be adjusted for sensitivity and specificity which are standard measures in evaluating the performance of such tests. Before we examine the effect of sensitivity and specificity on the estimate of prevalence, it should be recognized that the experimental error associated with the study design needed to evaluate the ST should be minimized. This is because the magnitude of this error affects the process of medical decision-making. One way to achieve this goal is by providing precise definition of the condition which the test is intended to detect. Intuitively, a perfect test is positive in all patients with the disease and negative in all patients who are disease free. Usually, this test is referred to as the "gold standard" test. However, one should realize that most tests are not perfect. This means that a negative answer does not always indicate the absence of the disease, because false negative results may occur. On the other hand, a positive result does not always mean that disease is present, because false positive may occur. In what follows, we shall focus on the issue of diagnostic or ST evaluation when comparing its results to the gold standard.

We shall focus on the situation where we have only two outcomes for the screening or the diagnostic tests, disease (D) and no-disease ($\overline{D}$). Note that, from a statistical point of view, evaluating the performance of a diagnostic test is the same as for STs, even though screening is a public health issue, while diagnostic is a clinical issue. Therefore, in this section, we will not distinguish diagnostic from screening when we are in the process of evaluating their performance.

1.2.8.2 Estimating Prevalence

The purpose of an ST is to determine whether a person belongs to the class (D) of people who have a specific disease. The test result indicating that a person is a member of this class will be denoted by T, and $\overline{T}$ for those who are nonmembers. The sensitivity of the test is $\eta = P(T|D)$, which describes the probability that a person with the disease is correctly diagnosed; the specificity of the test is $\theta = P(\overline{T}|\overline{D})$, which is the probability of a disease-free person being correctly diagnosed.

Let $\pi = P(D)$ denote the prevalence of the disease in the population tested. The results of an ST can be summarized in terms of θ, η, and π as in Table 1.2.

If we let $P(T) = p$ denote the apparent prevalence, then from the above table we have

$$p = \eta\pi + (1 - \theta)(1 - \pi). \tag{1.38}$$

Therefore,

$$1 - p = \pi(1 - \eta) + \theta(1 - \pi).$$

In mass screening programs the prevalence of π needs to be estimated. If a random sample of n individuals are tested, then we estimate p by the proportion ($\hat{P}$) of those who are classified in T.

Solving Equation 1.18 for π, we get the estimate of prevalence given by Rogan and Gladen (1978) as

$$\hat{\pi} = \frac{\hat{p} - (1 - \theta)}{\eta + \theta - 1}, \tag{1.39}$$

when θ and η are known. If η and θ are not known but instead are estimated from independent experiments that involved n_1 and n_2 individuals respectively, then π is estimated by

$$\hat{\pi}_2 = \frac{\hat{p} + \hat{\theta} - 1}{\hat{\eta} + \hat{\theta} - 1}. \tag{1.40}$$

TABLE 1.2

Layout of the Results of a Screening Test

Disease Status	Test Result		Total
	T	$\overline{T}$	
D	$\eta\pi$	$\pi(1 - \eta)$	π
$\overline{D}$	$(1 - \theta)(1 - \pi)$	$\theta(1 - \pi)$	$1 - \pi$
Total	p	$1 - p$	1

1.2.8.3 Estimating Predictive Value Positive and Predictive Value Negative

An important measure of performance of an ST, in addition to η and θ, is the predictive value of a positive test, $C = (PV+)$, or $P(D|T)$. From Table 1.2, we have

$$C = \frac{P(D \cap T)}{P(T)} = \frac{\eta \pi}{p}.$$

When η and θ are known, π is replaced by its estimate to get an estimate $\hat{C}$, for $P(D|T)$

$$\hat{C} = \frac{\hat{\pi}\eta}{\hat{p}} = \frac{\eta}{\eta + \theta - 1}\left[1 - \frac{1-\theta}{\hat{p}}\right].$$

As an exercise, we shall show that the asymptotic variance of $\hat{C}$ in this case is

$$\text{var}(\hat{C}) = \left[\frac{\eta(1-\theta)}{n + \theta - 1}\right]^2 \frac{(1-p)}{np^3}. \tag{1.41}$$

When η and θ are not known, but are estimated by $\hat{\eta}$ and $\hat{\theta}$ based on samples of sizes n_1 and n_2 where the ST is used on persons whose disease status is known, we replace η and θ by these estimated values such that an estimate of C is given by

$$\hat{C}_1 = \frac{\hat{\eta}}{\hat{\eta} + \hat{\theta} - 1}\left[1 - \frac{1-\hat{\theta}}{\hat{p}}\right]. \tag{1.42}$$

Gastwirth (1987) showed that as n, n_1, and n_2 increase, $\hat{C}_1 \sim N(C, \text{var}(\hat{C}_1))$.

1.2.8.4 Agreement and Performance of Diagnostic Tests

Sensitivity and specificity are basic measures of performance for a diagnostic test. Together, they describe how well a test can determine whether a specific disease or condition is present or absent. Each provides distinct and equally important information, and should be presented together:

Sensitivity is how often the test is positive when the disease is present.

Specificity is how often the test is negative when the disease is absent.

Usually, to estimate sensitivity and specificity, the outcome of the new test is compared to the true diagnosis using specimens from patients who are

TABLE 1.3

Results Comparing a New Test to True Diagnosis

		True Diagnosis	
		(+)	(−)
New Test	(+)	a	b
	(−)	c	d
Total		$a + c$	$b + d$

representatives of the intended use (both diseased and nondiseased) populations. Results are typically reported in a 2×2 table such as Table 1.3.

The new test has two possible outcomes, positive (+) or negative (−). Diseased patients are indicated as positive (+) true diagnosis, and nondiseased patients are indicated as negative (−) true diagnosis.

From Table 1.3, estimated sensitivity is the proportion of diseased individuals that are New Test +. Estimated specificity is the proportion of nondiseased individuals that are New Test −. The formulas are as follows:

$$\text{Estimated sensitivity} = 100\% \, (a/(a + c))$$

$$\text{Estimated sensitivity} = 100\% \, (d/(b + d)).$$

Here is an example of this calculation. Suppose one specimen is taken from each of the 220 patients in the target population. Each specimen is tested by the new test, and the diagnosis for each patient is determined. Fifty-one patients have the disease and 169 do not. The results are presented in a 2×2 table format in Table 1.4.

From Table 1.4, estimated sensitivity and specificity are calculated in the following manner:

$$\text{Estimated sensitivity} = 100\% \, (88/102) = 86.3\%$$

$$\text{Estimated sensitivity} = 100\% \, (336/338) = 99.4\%.$$

TABLE 1.4

Example of Results Comparing a New Test to True
Diagnosis for 220 Patients

		True Diagnosis		
		+	−	Total
New	+	88	2	90
Test	−	14	336	350
Total		102	338	440

Other quantities can be computed from this 2 × 2 table, too. These include positive predictive value, negative predictive value, and the positive and negative likelihood ratios. These quantities provide useful insight into how to interpret test results.

1.2.8.5 Calculating an Estimate of Agreement

When a new test is compared to an imperfect standard rather than clinical diagnosis or to a perfect standard, the usual calculations from the 2 × 2 table, $a/(a + c)$ and $d/(b + d)$, respectively, are biased estimates of sensitivity and specificity because the imperfect standard is not always correct. In addition, quantities such as positive predictive value, negative predictive value, and the positive and negative likelihood ratios cannot be computed since truth is unknown. However, being able to describe how often a new test agrees with an imperfect standard may be useful. To do this, a group of individuals (or specimens from individuals) are tested twice, once with the new test and once with the imperfect standard. The results are compared and can be reported in a 2 × 2 table as in Table 1.5.

The difference between Table 1.5 and Table 1.3 is that the columns of Table 1.5 do not represent truth, so data from Table 1.5 cannot be interpreted in the same way as Table 1.3. Data from Table 1.3 provides information on how often the new test is correct, whereas data from Table 1.5 provides information on how often the new test agrees with an imperfect standard.

From Table 1.5, one can compute several different statistical measures of agreement. A discussion by Shoukri (1999, 2000) on different types of agreement measures can be found under "Agreement, Measurement of" in the *Encyclopedia of Biostatistics and Encyclopedia of Epidemiology*. Two commonly used measures are the overall percent agreement and Cohen's kappa. The simplest measure is overall percent agreement. Overall percent agreement is the proportion of total specimens where the new test and the imperfect standard agree. One can calculate the estimated overall percent agreement from Table 1.5 in the following way:

$$\text{Overall percent agreement} = \frac{100\% \times (a + d)}{(a + b + c + d)}.$$

TABLE 1.5

Results Comparing a New Test to an Imperfect Standard

		Imperfect Standard	
		+	−
New	+	a	b
Test	−	c	d
Total		$a + c$	$b + d$

Since agreement on the absence of disease does not provide direct information about agreement on presence of disease, it may be useful to report two additional measures of agreement.

$$\text{Agreement of new test with imperfect standard-positive} = 100\% \times \frac{a}{(a+c)}$$

$$\text{Agreement of new test with imperfect standard-negative} = 100\% \times \frac{d}{(b+d)}.$$

As an example, consider the same 440 individuals as before. After all 440 are tested with both the new test and the imperfect standard, we have the following results.

From Table 1.6, calculate the agreement measures as follows.

$$\text{Overall percent agreement} = 100\% \times \frac{(80+342)}{440} = 95.9\%$$

$$\text{Agreement of new test with imperfect standard-positive} = 100\% \times \frac{80}{88} = 90.9\%$$

$$\text{Agreement of new test with imperfect standard-positive} = 100\% \times \frac{342}{352} = 97.2\%.$$

From Table 1.6, it is noted that the imperfect standard did not correctly classify all 440 patients. The imperfect standard classified 88 patients as positive and 352 as negative. From Table 1.4, in truth, 102 patients are diseased and 338 are nondiseased. Since the imperfect standard is wrong sometimes, one cannot calculate unbiased estimates of sensitivity and specificity from Table 1.6; however, one can calculate the agreement.

There are two major disadvantages with any agreement measure. One disadvantage is that "agreement" does not mean "correct." The other is that the

TABLE 1.6

Example of Results Comparing a New Test to an Imperfect Standard for 220 Patients

		Imperfect Standard		
		+	−	Total
New	+	80	10	90
Test	−	8	342	350
Total		88	352	440

agreement changes depend on the disease prevalence. The issue is discussed in more detail in Chapter 5.

1.2.8.6 *Estimation of Measures of Performance in Double Samples*

In the previous section, we estimated the prevalence using a ST when the true status of an individual is unknown. If however the true disease status of an individual can be determined by a test or device where it is not subject to misclassification, then we have what is traditionally known as the "gold standard." When there is no gold standard, then estimates of sensitivity and specificity should be replaced with an appropriate measure of agreement.

In most practical situations though, the ST is a relatively inexpensive procedure having less than perfect sensitivity and specificity; thus it tends to misclassify individuals. Using only the ST on all the n individuals results in a biased estimate of π as well. Because the use of the gold standard may be very costly due to the requirement for n_1 and n_2 additional individuals, the estimation of η and θ by this means may not be easily obtained.

To compromise between the two extremes, a double sampling scheme (DSS) was proposed by Tenenbein (1970). This DSS requires that:

1. A random sample of n individuals is drawn from the target population.
2. A subsample of n_1 units is drawn from n and each of these n_1 units is classified by both the ST and the gold standard.
3. The remaining $n - n_1$ individuals are classified only by the ST. Let x denote the number of individuals whose ST classification is diseased and let y denote the number of individuals whose ST classification is not diseased.

Using Tenenbein's notation, the resulting data can be presented as shown in Table 1.7.

The likelihood function of n_{ij} $(n_{11}, n_{10}, n_{01}, n_{00})'$ is proportional to

$$L(n_{ij}) = (\eta\pi)^{n_{11}} (\pi(1 - \eta))^{n_{10}} ((1 - \theta)(1 - \pi))^{n_{01}} (\theta(1 - \pi))^{n_0}.$$

TABLE 1.7

Data Layout for Tenenbein's Double Sampling Scheme

		s	S	
Gold Standard	D	n_{11}	n_{10}	$n_{1\cdot}$
	$\overline{D}$	n_{01}	n_{00}	$n_{0\cdot}$
		n_1	n_0	n_1
		X	Y	$n - n_1$

Conditional on n_{ij}, x has binomial distribution Bin $(p, n - n_1)$ and the likelihood of the experiment is proportional to

$$L = L(n_{ij}) \text{ Bin } (p, n - n_1)$$

$$\propto (\eta\pi)^{n_{11}} (\pi(1 - \eta))^{n_{10}} ((1 - \theta)(1 - \pi))^{n_{01}}$$

$$\times (\theta(1 - \pi))^{n_{00}} p^x (1 - p)^y.$$

Since

$$C = \frac{\eta\pi}{p}, \quad F = \frac{\pi(1 - \eta)}{1 - p},$$

then

$$1 - C = \frac{(1 - \theta)(1 - \pi)}{p}$$

and

$$1 - F = \frac{\theta(1 - \pi)}{1 - p}.$$

Writing L in terms of C, F, and P, we get

$$L \propto C^{n_{11}} (1 - C)^{n_{01}} (1 - F)^{n_{00}} F^{n_{10}} P^{x + n_1} (1 - P)^{y + n_0}.$$

Differentiating the logarithm of L with respect to C, F, and P, we get their maximum likelihood estimates (MLEs) as

$$\hat{C} = \frac{n_{11}}{n_1}, \quad \hat{F} = \frac{n_{10}}{n_{.0}}, \quad \text{and} \quad \hat{P} = \frac{x + n_{.1}}{x + y + n_1}.$$

Therefore, the MLE of π, η, and θ are, respectively, given as

$$\hat{\pi} = \frac{n_{11}}{n_{.1}} \frac{x + n_{.1}}{x + y + n_1} + \frac{n_{10}}{n_{.0}} \frac{y + n_{.0}}{x + y + n_1}$$

$$\hat{\eta} = 1 - \frac{\frac{n_{10}}{n_{.0}} \frac{y + n_{.0}}{x + y + n_1}}{\hat{\pi}},$$

and

$$\hat{\theta} = 1 - \frac{\dfrac{n_{01}}{n_{.1}} \dfrac{x + n_{.1}}{x + y + n_1}}{1 - \hat{\pi}}.$$

Note that $x + y + n_1 = n$.

The variance–covariance matrix of the MLEs is obtained by inverting Fisher's information matrix whose diagonal elements are given by

$$-E\left[\frac{\partial^2 \log L}{\partial C^2}\right], \quad -E\left[\frac{\partial^2 \log L}{\partial F^2}\right], \quad -E\left[\frac{\partial^2 \log L}{\partial p^2}\right],$$

and where

$$-E\left[\frac{\partial^2 \log L}{\partial C \, \partial F}\right] = -E\left[\frac{\partial^2 \log L}{\partial C \, \partial p}\right] = -E\left[\frac{\partial^2 \log L}{\partial F \, \partial p}\right] = 0$$

$$\frac{\partial^2 \log L}{\partial C^2} = \frac{n_{11}}{C^2} - \frac{n_{01}}{(1 - C)^2}$$

$$-E\left[\frac{\partial^2 \log L}{\partial C^2}\right] = \frac{1}{C^2} E(n_{11}) + \frac{1}{(1 - C)^2} E(n_{01})$$

$$= \frac{n_1 \eta \pi}{C^2} + \frac{n_1 (1 - \theta)(1 - \pi)}{(1 - C)^2}.$$

Since

$$C = \frac{\eta \pi}{p} \quad \text{and} \quad (1 - \theta)(1 - \pi) = p(1 - C),$$

then

$$-E\left[\frac{\partial^2 \log L}{\partial C^2}\right] = \frac{n_1 p}{C} + \frac{n_1 p}{1 - C} = \frac{n_1 p}{C(1 - C)}.$$

Moreover,

$$-E\left[\frac{\partial^2 \log L}{\partial F^2}\right] = \frac{1}{F^2} E(n_{10}) + \frac{1}{(1 - F)^2} E(n_{00})$$

$$= n_1 \frac{\pi(1 - \eta)}{F^2} + n_1 \frac{\theta(1 - \pi)}{(1 - F)^2}.$$

Since

$$F = \frac{\pi(1-\eta)}{1-P}, \quad \text{and} \quad (1-P)(1-F) = \theta(1-\pi),$$

then

$$-E\left[\frac{\partial^2 \log L}{\partial F^2}\right] = n_1 \frac{1-p}{F} + n_1 \frac{1-p}{1-F}$$

$$= \frac{n_1(1-p)}{F(1-F)}.$$

Finally,

$$-E\left(\frac{\partial^2 \log L}{\partial p^2}\right) = \frac{1}{p^2} E(x + n_{\cdot 1}) + \frac{1}{(1-p)^2} E(y + n_{\cdot 0})$$

$$= \frac{1}{p^2}\left[(n - n_1)p + n_1 p\right]$$

$$+ \frac{1}{(1-p)^2}\left[(n - n_1)(1 - p) + (1 - p)n_{\cdot 1}\right]$$

$$= \frac{n}{p(1-p)}.$$

As n_1 and n increase,

$$\text{var}(\hat{C}) = \frac{C(1-C)}{n_1 p},$$

$$\text{var}(\hat{F}) = \frac{F(1-F)}{n_1(1-p)},$$

$$\text{var}(\hat{p}) = \frac{p(1-p)}{n}.$$

To find the variance of $\hat{\pi}$, we note that

$$\hat{\pi} = \hat{C}\,\hat{p} + \hat{F}(1-\hat{p}).$$

Using the DM, we have to evaluate the first order of approximation

$$\text{var}(\hat{\pi}) = \text{var}(\hat{C})\left(\frac{\partial \pi}{\partial C}\right)^2 + \text{var}(\hat{F})\left(\frac{\partial \pi}{\partial F}\right)^2 + \text{var}(\hat{p})\left(\frac{\partial \pi}{\partial p}\right)^2,$$

where the partial derivatives are evaluated at the true values of the parameters. Hence,

$$\text{var}(\hat{\pi}) = \frac{C(1-C)}{n_1}p + \frac{F(1-F)}{n_1}(1-p) + \frac{p(1-p)}{n}(C-F)^2.$$

Using the following identities by Tenenbein (1970)

$$\pi\eta(1-\eta) + (1-\pi)\theta(1-\theta) \equiv p(1-p) - \pi(1-\pi)(\eta+\theta-1)^2,$$

$$C(1-C)p + F(1-F)(1-p) \equiv \pi(1-\pi)\left[1 - \frac{\pi(1-\pi)}{p}(1-p)(\eta+\theta-1)^2\right],$$

and

$$p(1-p)(C-F)^2 \equiv \frac{\pi^2(1-\pi)^2}{p(1-p)}(\eta+\theta-1)^2,$$

we have

$$\text{var}(\hat{\pi}) = \frac{\pi(1-\pi)}{n_1}\left[1 - \frac{\pi(1-\pi)}{p(1-p)}(\eta+\theta-1)^2\right] + \frac{\pi^2(1-\pi)^2}{np(1-p)}(\eta+\theta-1)^2. \quad (1.43)$$

For two dichotomous random variables whose joint distribution is given as in Table 1.2, the cross-correlation is given by

$$\lambda = \frac{P(D\cap T) - P(T)P(D)}{\sqrt{P(D)(1-P(D))P(T)(1-P(T))}}$$

$$= \frac{\eta\pi - p\pi}{\sqrt{\pi(1-\pi)p(1-p)}} = \frac{\pi(\eta-p)}{\sqrt{\pi(1-\pi)p(1-p)}}.$$

Since

$$\eta - p = (1-\pi)(\eta+\theta-1)$$

then

$$\lambda = \sqrt{\frac{\pi(1-\pi)}{p(1-p)}}(\eta + \theta - 1).$$

Tenenbein (1970) defined the coefficient of reliability λ^2 as a measure of the strength between the ST and the gold standard. From Equation 1.43 we get

$$\text{var}(\hat{\pi}) = \frac{\pi(1-\pi)}{n_1}(1-R) + \frac{\pi(1-\pi)}{n}R.$$

It turns out that the asymptotic variance of $\hat{\pi}$ is the weighted average of the variance of a binomial estimate of π based on n_1 measurements from the gold standard and the variance of a binomial estimate of π based on n measurements from the gold standard. One should also note that:

1. When $\lambda = 0$, that is when the standard test (ST) is useless, then the precision of $\hat{\pi}$ depends entirely on the n_1 measurements from the gold standard.

2. When $\lambda = 1$, that is when the ST is as good as the gold standard, the precision of $\hat{\pi}$ depends on the n measurements from the gold standard. This is because in using n measurements from the ST, we get the same precision as a binomial estimate of π based on n measurements from the gold standard.

Example 1.1

Mastitis is a frequently occurring disease in dairy cows. In most cases, it is caused by bacteria.

To treat a cow it is quite important to identify the specific bacteria causing the infection. Bacterial culture is the gold standard but it is a time-consuming and expensive procedure. A new test of interest, pro-Staph ELISA procedure, is cheap and easily identifies the *Staphylococcus aureus* bacteria.

Milk samples were taken from 160 cows. Each sample was split into two halves, one half was tested by bacterial culture and the other half was tested by the ELISA. Further, milk samples taken from 240 cows were evaluated by the ELISA test. The data obtained are found in Table 1.8.

Here, $n_{11} = 23$, $n_{10} = 13$, $n_{01} = 18$, $n_{00} = 106$, $x = 62$, and $y = 178$. The sample sizes are $n_1 = 160$ and $n = 400$.

$$\hat{\pi} = \frac{23}{41}\left(\frac{62+41}{400}\right) + \frac{13}{119}\left(\frac{178+119}{400}\right) = 0.226$$

$$\hat{\eta} = 1 - \frac{\dfrac{13}{119}\left(\dfrac{178+119}{400}\right)}{0.226} = 1 - \frac{0.0811}{0.226} = 0.641$$

TABLE 1.8

Test Outcomes for Bacterial Identification using ELISA and a "New Test"

		New Test		
		1	0	Total
ELISA	1	23	13	36
	0	18	106	124
	Total	41	119	160
		62	178	240

$$\hat{\theta} = 1 - \frac{\frac{18}{41}\left(\frac{62 + 41}{400}\right)}{0.774} = 0.854$$

$$\hat{p} = (0.641)(0.226) + (1 - 0.854)(1 - 0.226) = 0.258$$

$$\hat{R} = \frac{(0.226)(1 - 0.226)}{0.258(1 - 0.258)}(0.641 + 0.854 - 1)^2 = 0.224$$

$$\text{var}(\hat{\pi}) = \frac{0.226(1 - 0.226)}{160}(1 - 0.224) + \frac{0.226(1 - 0.226)}{400}(0.224)$$

$$= 0.00085 + 0.000095 = 0.00095.$$

A 95% confidence interval on π is (0.166, 0.286). These results indicate that the true prevalence lies between 17% and 29%.

EXERCISES

E1.1 Suppose that (x_1, x_2) has a bivariate normal distribution. Define how the new random variables $S = x_1 + x_2$ and $D = x_1 - x_2$.
   a. Find the mean of variances of (S, D).
   b. Use the results of E1.1.a to derive $E(D/S)$ and $\text{var}(D/S)$

E1.2 Show, from the results of Bernoulli trials that, $\text{var}(\hat{\pi})$, the estimated prevalence given in Equation 1.39 is $\text{var}(\hat{\pi}) = (P(1 - P)/n(\eta + \theta - 1))^2$.

E1.3 Derive the variances $\hat{C}$ and $\hat{F}$.

E1.4 Let (y_1, y_2) denote the outcome of a Bernoulli trial, conditional on the probability of success $P_r(y_i = 1 \mid p) = p$, with $n = 2$. Suppose that the probability of success p has β-distribution with the transformed parameters, $\rho = (1 + a + b)^{-1}$, $\pi = a(a + b)^{-1}$. Write down the four probabilities of the β-binomial distribution. Evaluate the correlation between (y_1, y_2).

E1.5 In Exercise E1.4, derive the unconditional mean and the unconditional variance of the sum $S = y_1 + y_2$ in terms of π and ρ.

E1.6 Let $(y_1, y_2, ..., y_n)$ be the outcomes of independent Bernoulli trials with $P_r(y_i = 1|p) = p$. Assume that the probability of success p has β-distribution with the transformed parameters $\rho = (1 + a + b)^{-1}$, $\pi = a(a + b)^{-1}$. Based on the β-binomial outcomes, find the moment estimators of (π, ρ).

2

Reliability for Continuous Scale Measurements

2.1 Introduction

We begin this chapter by defining what is meant by error, accuracy, precision, repeatability, and reliability. Analytical error is the sum of two error components: random error (imprecision) and systematic error (inaccuracy). Random error is the amount of variability inherent in the measuring instrument, while systematic error is the deviation from the true state of nature. Random error may either be positive or negative relative to the observed mean value of replicate determinations (Koch and Peters, 1999). On the other hand, systematic error or bias may either be positive or negative.

The accuracy of a measuring technique is defined by the International Federation of Clinical Chemists (IFCC) as the closeness of the agreement between the measured value and the "true" value measured by a nonfaulty device. Definitive methods, such as mass spectrometry, are used to develop primary reference materials, which can then be used for development of reference methods by manufacturers. Comparative method means have been shown to closely approximate true values. Comparative method means, to a great extent, are obtained from the measurements generated by multiple laboratories using variety of instruments and techniques. On the other hand, precision of a measuring technique means its ability to reproduce its own results. Such measurement errors can seriously affect the interpretability of measurements or readings, no matter how they are obtained (by x-ray, ultrasound, interviews, or by laboratory assay). Efforts have been made by the International Organization of Standardization (IOS) to use the term accuracy to quantify both systematic and random errors. Systematic bias is measured by accuracy, while random error is measured by precision. Among applied statisticians, the definitions made by IOS are quite acceptable. Repeatability is the closeness of agreement between measurements taken under identical experimental conditions. Finally, reliability is the ability of the measuring devise to differentiate or discriminate between subjects. The intraclass correlation coefficient (ICC) has emerged as a universal and widely accepted reliability index (Bartko, 1966; Ebel, 1951; Haggard, 1958).

There are several versions of the ICC that can give quite different results when applied to the same set of measurements (Rousson et al., 2002). What should be emphasized is that (a) researchers need to be made aware of the differences between the forms and that each form is appropriate for a specific situation defined by the experimental design and the conceptual intent of the study, (b) the fundamental interpretation of an ICC is that it is a measure of the proportion of a variance that is attributable to objects of measurements. The objects might be students, slides, x-rays, or packages and the corresponding measurements might be readings, IQs of students, weights of newborns, or test score of students.

The first step in constructing an estimate of ICC is to specify an additive model that best describes the outcome of the experiment in a reliability study. The guidelines for choosing the appropriate form of the ICC call for one of the three possibilities outlined by Shrout and Fleiss (1979): (a) Is one-way or two-way analysis of variance (ANOVA) appropriate for the reliability study? (b) Do the differences between raters mean readings relevant to the reliability index? (c) Is the unit of analysis an individual rating or the mean of several ratings?

2.2 Models for Reliability Studies

In a typical interrater reliability study, each of the random samples of k subjects (items) is judged by n raters, and the layout is shown in Table 2.1. Several situations may produce this table:

Case 1. Each subject is rated by a different set of n raters, randomly selected from a larger population of raters, and in this case the ordering of j is irrelevant. This situation represents a nested design because unordered observations are nested within subjects. In such cases, the random effect one-way ANOVA can be used to represent this data.

TABLE 2.1

One-Way Data Layout Subjects

Measurements	1	2	i	k
1	y_{11}	y_{21}	y_{i1}	y_{k1}
2	y_{12}	y_{22}	y_{i2}	y_{k2}
$\vdots$				
j	y_{1j}	y_{2j}	y_{ij}	y_{kj}
$\vdots$				
n	y_{1n}	y_{2n}	y_{in}	y_{kn}

Case 2. A random sample of n raters is selected from a large population, and each rater rates each subject, that is, each rater rates k subjects.

Case 3. Each subject is rated by each of the same n raters who are the only raters of interest.

Under Case 1, let y_{ij} denote the jth rating $(j = 1, 2, ..., n)$ on the ith subject $(i = 1, 2, ..., k)$.

We assume the following linear model for y_{ij}:

$$y_{ij} = \mu + b_i + w_{ij}.$$

The underlying assumptions of this model are: μ (the population mean for all measurements) is constant; b_i is the deviation of the ith subject from μ; and w_{ij} is a residual component. The component b_i is assumed to vary normally with a mean of zero and a variance σ_b^2 and to be independent of all other components in the model. Moreover, it is assumed that the w_{ij} terms are normally and independently distributed with a mean of zero and a variance σ_w^2.

The expected mean-squares in the ANOVA table related to the one-way random effects model appear in Table 2.2.

From the expectations of the mean squares, one can see that the WMS is an unbiased estimate of σ_w^2, and that (BMS – WMS)/n is an unbiased estimate of σ_b^2. The classical definition of the ICC (ρ) is

$$\rho = \frac{\sigma_b^2}{\sigma_b^2 + \sigma_w^2}. \tag{2.1}$$

It is, therefore, the ratio of intersubject variability to total variability. The estimate of ρ then, takes the form

$$\hat{\rho}_1 = \frac{BMS - WMS}{BMS + (n-1)WMS}. \tag{2.2}$$

The estimate of ρ_1 is consistent, but biased.

If the reliability study has the design of Case 2 or Case 3, a two-way model can be used to represent the data. The reason is that there is a systematic source of variation between subjects and between raters. For example, if the

TABLE 2.2

ANOVA and Mean-Square Expectations for the One-Way Random Effects Model

Source of Variation	df	MS	EMS
Between subjects	$k - 1$	BMS	$\sigma_w^2 + n\sigma_b^2$
Within subjects	$k(n-1)$	WMS	σ_w^2

rows in Table 2.1 represent different scales used to weigh chemical compounds, then the scales may differ in their sensitivities, thus creating a separable source of variation. In this and similar situations, we specify a two-way model, and from a design point of view a randomized blocks design in which the column variable (subjects) is crossed with the row variable.

The two-way model for Cases 2 and 3 differ from the one-way model; in that the components of w_{ij} are further specified. Moreover, there is only one ICC for the one-way data whereas, there are four ICCs given for two-way data. These distinctions among models are important because they have implications for the ICCs that can be calculated and for their interpretations.

Now, since the same n raters rate all k subjects, the component representing the jth rater's effect may be estimated. The model equation representing this situation is given as:

$$y_{ij} = \mu + b_i + r_j + (br)_{ij} + e_{ij}. \tag{2.3}$$

It is appropriate for both Cases 2 and 3. In Equation 2.3, the terms y_{ij}, μ, and b_i are defined as in Equation 2.1; r_j is the deviation from the mean of the jth rater's score; $(br)_{ij}$ is the degree to which the jth rater deviates from its usual rating when presented to the ith subject (interaction effect); and e_{ij} is the random error in the jth rater's scoring of the ith subject. In both Cases 2 and 3 the subject effect b_i is assumed to vary normally with a mean of zero and variance σ_b^2 (as in Case 1), and the error terms e_{ij} are assumed to be independently and normally distributed with a mean zero and variance σ_e^2.

It should be noted that Case 2 differs from Case 3 with respect to the assumption made on r_j and $(br)_{ij}$ in Equation 2.3. In Case 2, r_j is a random variable assumed to be normally distributed with mean zero and variance σ_e^2; under Case 3, it is a fixed effect with the constraints

$$\sum_{j=1}^{n} r_j = 0.$$

It is well known from the principles of experimental design that in the absence of replication, the term $(br)_{ij}$ is not estimable. Nevertheless, it must be kept in the model because the properties of the interaction are different in the two cases.

This leads us to consider the following cases separately: Case 2, with r_j being random, we consider the case when $(br)_{ij} = 0$ (i.e., interaction is absent. Table 2.3 gives the ANOVA mean square errors and their expectations). We denote the case when interaction is absent by Case 2-A.

The notations in the third column are: BMS = between-patients mean-square; RMS = between-raters mean-square; MSE = mean-square error; WMS = within mean-square.

The ANOVA table for Case 2-A is identical to the ANOVA Table 2.3, if we set $\sigma_{be}^2 = 0$.

TABLE 2.3

Two-Way Random Effects Model with Interaction: Case 2

Source of Variation	df	Mean-Squares	EMS
Between subjects	$k-1$	BMS	$n\sigma_b^2 + \sigma_{br}^2 + \sigma_e^2$
Within subjects	$k(n-1)$	WMS	$\sigma_r^2 + \sigma_{br}^2 + \sigma_e^2$
Between raters error	$n-1$	RMS	$k\sigma_r^2 + \sigma_{br}^2 + \sigma_r^2$
Error	$(n-1)(k-1)$	MSE	$\sigma_{br}^2 + \sigma_e^2$

The ANOVA estimator of the ICC under Cases 2 and 2-A is given by

$$\hat{\rho}_2 = \frac{BMS - MSE}{BMS + (n-1)MSE + n(RMS - MSE)/k}.$$

Rajaratnam (1960) and Bartko (1976) were the first to give the above expression. Because in Case 3 raters are considered fixed, the statistical model used to construct the population, ICC is different from Case 2. In this case, the ANOVA in Table 2.4 provides the appropriate sources of variation and the expected mean square (EMS), for the mixed model (Equation 2.4) that describes Case 3:

$$Y_{ij} = \mu + b_i + r_j + (br)_{ij} + e_{ij}. \tag{2.4}$$

Here, the r_i's are fixed so that

$$\sum_{j=1}^{n} r_j = 0, \quad \sum_{j=1}^{n} (br)_{ij} = 0,$$

TABLE 2.4

ANOVA for the Mixed Effects Model in Equation 2.4 for Case 3

Source of Variation	df	Mean-Squares	EMS
Between subjects	$k-1$	BMS	$n\sigma_b^2 + \sigma_e^2$
Within subjects	$k(n-1)$	WMS	$\Phi_r^2 + \dfrac{n}{n-1}\sigma_{br}^2 + \sigma_e^2$
Between raters error	$(n-1)$	RMS	$k\Phi_r^2 + \dfrac{n}{n-1}\sigma_{br}^2 + \sigma_e^2$
	$(n-1)(k-1)$	MSE	$\dfrac{n}{n-1}\sigma_{br}^2 + \sigma_e^2$

and the parameter corresponding to σ_r^2 in Case 2 is

$$\phi_r^2 = \frac{1}{n-1}\sum_{j=1}^{n} r_j^2.$$

When interaction is absent, we set $(br)_{ij} = 0$. In Case 3, in the absence of interaction, we set $\sigma_{br}^2 = 0$.

We should note that one implication of the rater being fixed is that no unbiased estimator of σ_b^2 is available when $\sigma_b^2 > 0$. On the other hand, σ_b^2 is no longer the covariance between (y_{ij}, y_{il}) $(j \neq 1)$. The interaction term has variance σ_{br}^2, and

$$\text{cov}\left(y_{ij}, y_{il}\right) = \sigma_b^2 - \frac{\sigma_{br}^2}{n-1},$$

and therefore, under Case 3 the ICC is given as

$$\rho_3 = \frac{\sigma_b^2 - \sigma_{br}^2/(n-1)}{\sigma_b^2 + \sigma_{br}^2 + \sigma_e^2},$$

which is consistently estimated by

$$\hat{\rho}_3 = \frac{BMS - MSE}{BMS + (n-1)MSE}.$$

Example 2.1

Three scales are used to weigh 10 packets. This is an example of one and two-way random and mixed effects model. The main objective is to calculate the intraclass correlation as an index of reliability.

The first model, is a one-way random effects model which assumes that there is no scale effect.

The second model assumes that the scales are fixed, while the last model assumes that the scales are a random sample of a large population of scales (raters). The SS code is given below

```
data weight;
input packet scale$ weight;
cards;
/* One-way Random effect model*/
    proc glm data = sbp;
        class packet;
        model weight = packet;
        random packet;
        run;

        /* Two-way mixed effects model */
```

```
proc glm data = sbp;
class packet scale;
model weight = packet scale;
random packet;
run;

/* Two-way random effects model */
proc glm data = sbp;
class packet scale;
model weight = packet scale;
random packet scale;
run;
```

Summary Statistics

R-Square	C.V.	Root MSE	WEIGHT Mean		
0.9997	1.046872	8.517	813.53		
Source	**DF**	**Type III SS**	**Mean-Square**	**F Value**	**Pr > F**
PACKET	9	5992180.800	665797.86666667	9179.20	0.0001

Note: The ICC from the one-way random effects model is calculated as follows: The within-packet variance = 72.53, the between-packets variance = (665797.87 − 72.53)/ 3 = 221908. Hence, the ICC $\hat{\rho}_1 = 221908/(221908 + 72.53) = 0.999$

Source	DF	Type III SS	Mean-Square	F Value	Pr > F
PACKET	9	5992180.800	665797.866	7128	0.0001
SCALE	2	517.066	258.5333	4.98	0.018

Note: The ICC under the two-way mixed effects model is calculated as follows: The within variance is 51.87, the between variance = (665797 − 51087)/3 = 221915. ICC $\hat{\rho}_2 = 221915/ (221915 + 51.87) = 0.999$.

Source	DF	Type III SS	Mean-Square	F Value	Pr > F
PACKET	9	5992180.8	665797.87	12836.72	0.0001
SCALE	2	517.067	258.533	4.9	0.0189

Note: The ICC under the two-way random effects model is calculated as follows: Within variance = 51.87, the between-packets variance = (665797.87 − 51.87)/3 = 221915. The between scales variance = (258.533 − 51.87)/10 = 20.67. ICC $\hat{\rho}_3 = 221915/(221915 + 20.67 + 51.87) = 0.999$.

Remark

It is noted that the value of ICC under the three models is the same. Generally, this is not the case.

2.2.1 Inference Procedures on the Index of Reliability for Case 1

2.2.1.1 Interval Estimation

We start this section by deriving the large sample variance of ρ_1 from Case 1. Since the point estimator $\hat{\rho}_1$ is obtained from the one-way ANOVA as

$$\hat{\rho}_1 = \frac{BMS - WMS}{BMS + (n-1)WMS} = \frac{F-1}{F+(n-1)},$$

where $F = BMS/WMS$ is the usual variance ratio statistic. The large sample variance of $\hat{\rho}_1$ derived by Fisher (1925) under the assumption of normality is to the first order of approximation, given by

$$V(\hat{\rho}_1) = \frac{2(1-\rho)^2(1+(n-1)\rho)^2}{kn(n-1)}. \qquad (2.5)$$

Because of their computational simplicity, approximate confidence limits on ρ_1 may be obtained. An approximate $(1 - \alpha)$ 100% confidence interval for ρ, is given by

$$\frac{(F/F_u)-1}{n-1+F/F_u}, \quad \frac{(F/F_L)-1}{n-1+F/F_L} \qquad (2.6)$$

where $P_r\{F_L \le F(k-1, k(n-1)) \le F_u\} = 1 - \alpha$.

Donner and Wells (1986) showed that an accurate approximation to the exact limits is given, for moderately large value of k, by

$$\left\{\hat{\rho}_1 - Z_\alpha\sqrt{\hat{V}(\hat{\rho}_1)}, \hat{\rho}_1 + Z_\alpha\sqrt{\hat{V}(\hat{\rho}_1)}\right\} \qquad (2.7)$$

where $V(\hat{\rho}_1)$ is defined by Equation 2.5, with $\hat{\rho}_1$ replacing ρ, and Z_α is the two-sided critical value of the standard normal distribution corresponding to α. Unlike the approximation given by Equation 2.6, the limits given by Equation 2.7 are accurate over a broad range of values for ρ. For the given example, the estimated standard error of $\hat{\rho}_1$ is 0.0005. Therefore, the 95% confidence interval on ρ_1 using Equation 2.7 is 0.998, 1.00.

2.2.1.2 Hypothesis Testing

2.2.1.2.1 Exact Methods

Suppose that we are interested in testing the hypothesis $H_0 : \rho \le \rho_0$ versus $H_1 : \rho > \rho_0$ for preselected Type I error rate α and power $1 - \beta$. Note that $\rho_0 = 0$ is of no practical use in reliability studies. Landis and Koch (1977) provided benchmarks for values of reliability coefficient (0–0.20) as poor, (0.21–0.40) as fair, (0.1–0.60) as moderate, (0.71–0.80) substantial, and (0.81–1.00) as perfect. For example, if we are to demonstrate substantial level of reliability the hypotheses are $H_0 : \rho = 0.60$ and $H_1 : \rho > 0.60$. From Scheffe (1959), the null hypothesis is rejected if (BMS/WMS) exceeds C, $F_{\alpha;v_1 v_2}$, where $C = 1 + (n\rho_0/1 - \rho_0)$ and $F_{\alpha;v_1 v_2}$ is the tabular value of F with v_1 and v_2 degrees of freedom at the α level of significance, $v_1 = k - 1$, $v_2 = k(n-1)$. Moreover, the power is given by

$$1 - \beta = \Pr\left[\frac{DMS}{WMS} \geq C_0 F_{\alpha; v_1, v_2}\right]$$

where

$$C_0 = \left(1 + \frac{n\rho_0}{1 - \rho_0}\right) \Big/ \left(1 + \frac{n\rho}{1 - \rho}\right)$$

and $(BMS/WMS) = (1 + (n-1)\hat{\rho}_1/1 - \hat{\rho}_1)$ is obtained from the ANOVA Table 2.2.

2.2.1.2.2 Approximate Methods
Fisher (1925) showed that for large k

$$\hat{Z} = \frac{1}{2}\ln\left\{\frac{1 + (n-1)\hat{\rho}_1}{1 - \hat{\rho}_1}\right\} = \frac{1}{2}\ln(F)$$

has an approximate normal distribution with mean

$$\mu_Z = \frac{1}{2}\ln\left\{\frac{1 + (n-1)\rho_1}{1 - \rho_1}\right\}$$

and variance

$$v_Z = \frac{kn - 1}{2k(k-1)(n-1)}.$$

The inverse of this transformation is given by

$$I(Z) = \frac{\exp(2Z) - 1}{\exp(2Z) + n - 1}.$$

From Weinberg and Patel (1981), an accurate 95% approximate confidence interval for ρ is given by (L, U) where

$$L = I\left(\hat{Z} - 1.96\sqrt{v_Z}\right)$$

$$U = I\left(\hat{Z} + 1.96\sqrt{v_Z}\right).$$

For the data in the example we have

$$\hat{Z} = \frac{1}{2}\ln\left\{\frac{1 + (n-1)\hat{\rho}_1}{1 - \hat{\rho}_1}\right\} = \frac{1}{2}\ln(F) = 4,$$

$v_z = 0.08$. Therefore, the 95% confidence interval based on the Fisher's approximation (0.997, 0.9996). This is very close to the result obtained using Equation 2.7.

2.3 Testing the Equality of Two Independent ICCs

The well-known Fisher's (1925) approximation can be used to test the hypothesis $H_0 : \rho_1 = \rho_2$ versus $H_1 : \rho_1 \neq \rho_2$.

Let $\hat{\rho}_{1A}$ and $\hat{\rho}_{1B}$ be estimates of ICC from two independent samples. Fisher's result can be extended as follows:

For lth sample

$$Zl = \frac{1}{2}\ln\left\{\frac{1 + (n_l - 1)\hat{\rho}_{1l}}{1 - \hat{\rho}_{1l}}\right\} \quad l = A, B$$

has an approximate normal distribution with mean

$$Z(\rho_{1l}) = \frac{1}{2}\ln\left\{\frac{1 + (n_l - 1)\rho_{1l}}{1 - \rho_{1l}}\right\}$$

and variance

$$v_l = \frac{k_l n_l - 1}{2k_l(k_l - 1)(n_l - 1)}$$

where k_l is the number of subject in lth sample and n_l is the number of measurements on a subject from the lth sample ($l = A, B$).

Following Donner (1986), define $W_l = V_l^{-1}$, and

$$\bar{Z}_W = \frac{Z(\hat{\rho}_{1A})W_A + Z(\hat{\rho}_{1B})W_B}{W_A + W_B}.$$

Then an approximate test on the hypothesis $H_0 : \rho_{1A} = \rho_{2B}$ is obtained by referring

$$T = W_A\left(Z(\hat{\rho}_{1A}) - \bar{Z}_W\right)^2 + W_B\left(Z(\hat{\rho}_{1B}) - \bar{Z}_W\right)^2$$

to tables of chi-square distribution with one-degree-of-freedom.

2.4 Testing the Equality of Two Dependent ICCs

Methods for testing the quality of two or more independent ICC has extensively been discussed by many researchers (e.g., Donner and Bull, 1983; Konishi and Gupta, 1989; Mian and Shoukri, 1997; Paul and Barnwal, 1990). Under the assumption of multivariate normality, the likelihood-based approach was used and provided efficient procedure for comparing ICC. The problem of comparing dependent ICC has been recently investigated by Bowerman et al. (1990), who reported a study in which two radiologists and two clinical hematologists, independently evaluated the radiographic vertebral index on 40 radiographs from patients with myeloma. Comparing the expertise of the raters, leads to the comparison between two dependent ICCs.

Donner and Zou (2002) provided a comprehensive review and empirical evaluation of the statistical properties of procedures used to test the null hypothesis $H_0 : \rho_1 = \rho_2$. These methods take into account the correlation induced due to the fact the raters are repeatedly evaluating the same set of subjects. Typically, the data layout would be as shown in Table 2.5.

2.4.1 Methods for Testing $H_0 : \rho_1 = \rho_2$

2.4.1.1 Donner and Zou

Let r_1 be the ANOVA estimator of ρ_1 ($l = 1, 2$) from the one-way random effects model. Similar to the case of independent sampling we employ Fisher's (1925) transformation,

$$Z_l = \frac{1}{2}\ln\left\{\frac{1 + (n_1 - 1)r_1}{1 - r_1}\right\}$$

TABLE 2.5

Data Layout for Dependent Intraclass Correlations

Raters	Subjects				
	1	2	...	i	k
1	y_{111}	y_{211}		y_{i11}	y_{k11}
	y_{112}	y_{212}		y_{i12}	y_{k12}
	⋮	⋮		⋮	⋮
	y_{11n1}	y_{21n1}		y_{i1n1}	y_{k1n1}
2	y_{121}	y_{221}		y_{i21}	y_{k21}
	y_{122}	y_{222}		y_{i22}	y_{i22}
	⋮	⋮		⋮	⋮
	y_{12n2}	y_{22n2}		y_{i2n2}	y_{k2n2}

which has an approximate normal distribution with mean

$$\mu_l = \frac{1}{2}\ln\left\{\frac{1+(n_1-1)\rho_1}{1-\rho_1}\right\}$$

and variance

$$v_l = \frac{kn_l-1}{2k(k-1)(n_l-1)} \qquad l = 1,2.$$

Donner and Zou (2002) used the delta method and results by Elston (1975) to show that to the first order of approximation

$$\text{cov}(Z_1,Z_2) = \frac{n_1n_2\rho_{12}^2}{2k\left\{1+(n_1-1)\rho_1\right\}\left\{1+(n_2-1)\rho_2\right\}}.$$

Therefore, under $H_0 : \rho_1 = \rho_2 = \rho$ (unspecified)

$$\theta = E(Z_1 - Z_2) = \frac{1}{2}\ln\left\{\frac{1+(n_1-1)\rho}{1+(n_2-1)\rho}\right\}$$

and

$$V = \text{var}(Z_1 - Z_2) = V_1 + V_2 - 2\overline{\text{cov}}(Z_1, Z_2)$$

where $\overline{\text{cov}}(Z_1, Z_2)$ is equal to $\text{cov}(Z_1, Z_2)$ evaluated at $\rho_1 = \rho_2 = \rho$, and ρ_{12} is replaced by some estimator.

Therefore,

$$T = \frac{(Z_1 - Z_2) - (\hat{\theta})}{(\hat{V})^{1/2}}$$

is approximately distributed as a standard normal variable, and $\hat{\theta}$ is obtained on substituting the common ANOVA estimate r for ρ.

2.4.1.2 Alswalmeh and Feldt

Alswalmeh and Feldt (1994) developed a test statistic on the null hypothesis $H_0 : \rho_1 = \rho_2$. They proposed rejecting H_0 for extreme values of

$$T = \frac{1-\hat{\rho}_1}{1-\hat{\rho}_2}.$$

The statistics T is approximately distributed as an Γ random variable with d_1 and d_2, degrees of freedom, where

$$d_2 = \frac{2M}{M-1}, \quad d_1 = \frac{2d_2^3 - 4d_2^2}{(d_1 - 2)^2 (d_2 - 4)V - d_2^2},$$

$$M = \frac{E_1}{E_2} + \frac{E_1}{E_2^3}V_2 - \frac{C_{12}}{E_2^2},$$

$$E_j = \frac{v_j}{v_2 - 2} - \frac{(1 - \rho_j)}{(k-1)}, \quad j = 1, 2,$$

$$V_j = \frac{2v_j^2 (C_j + v_j - 2)}{C_j (v_j - 2)^2 (v_j - 4)} - \frac{2(1 - \rho_j)}{k-1},$$

$$v_j = \frac{2(k-1)}{1 + \rho_j^2}, \quad C_j = (k-1), \quad C_{12} = \frac{2}{(k-1)}\rho_{12}^2,$$

and

$$V = \left(\frac{E_1}{E_2}\right)^2 \left[\frac{V_1}{E_1^2} + \frac{V_2}{E_1^2} - \frac{2C_{12}}{E_1 E_2}\right].$$

Unfortunately, the test is only valid when $k > 50$.

The parameter ρ_{12} is the correlation between individual measurements x_j and y_j from the two methods. As a moment estimator of ρ_{12}, Ramasundarahet-tige et al. (2009) suggested the following procedure:

Step 1: Suppose that we have three measurements taken for the first subject by the first rater denoted by x_1, x_2, x_3 and another three measurements taken by the second rater on the same subject, denoted by y_1, y_2, y_3.

Step 2: For this subject form all the possible pairings: $(x_1, y_1),(x_1, y_2),$ $(x_1, y_3),(x_2, y_1),(x_2, y_2),(x_2, y_3),(x_3, y_1),(x_3, y_2),(x_3, y_3)$. Repeat the same procedure for all other subjects.

Step 3: Calculate the Pearson product moment correlation between the above pairs for all the subjects. The resulting correlation is a moment estimator of ρ_{12}.

Example 2.2

Accurate and reproducible quantification of brain lesion count and volume in multiple sclerosis (MS) patients using magnetic resonance imaging (MRI) is a vital tool for the evaluation of disease progression and patient response to therapy. Current standard methods for obtaining these data are largely manual and subjective and are therefore, error-prone and subject to inter- and intra-operator variability. Therefore, there is a need for a rapid automated lesion quantification method. Ashton et al. (2003) compared manual measurements and an automated data technique known as geometrically constrained region growth (GEORG) of the brain lesion volume of three MS patients, each measured 10 times by a single operator for each method. The data are presented in Table 2.6.

Data analysis for Example 2.2.
   The SAS code is:

```
ods rtf body = 'D:EXAMPLE_2.2.rtf';
Title 'Comparing two dependent ICCC';
/* ESTIMATION OF THE CORRELATION COEFFICIENT FOR EACH
   METHOD */
                proc sort data=coefvar;
                by method;
                proc glm data=coefvar;
                   by method;
                   class patient;
                   model y= patient;
                   random patient;
                   run;
```

The estimated cross correlation is $\rho_{12} = 0.996$.

The estimated correlation between patients variance component is 53, and the ICC for the G method is $\hat{\rho}_G = 0.999$. It can also be shown that $\hat{\rho}_M = 0.977$.

To test the null hypothesis $H_0 : \rho_1 = \rho_2 = $ unspecified quantity, using the above data we have

TABLE 2.6

Results for 10 Replicates on Each of Three Patient's Total Lesion Burden

Patient	Method[a]	1	2	3	4	5	6	7	8	9	10
1	M	20	21.2	20.8	20.6	20.2	19.1	21	20.4	19.2	19.2
	G	19.5	19.5	19.6	19.7	19.3	19.1	19.1	19.3	19.2	19.5
2	M	26.8	26.5	22.5	23.1	24.3	24.1	26	26.8	24.9	27.7
	G	22.1	21.9	22	22.1	21.9	21.8	21.7	21.7	21.7	21.8
3	M	9.6	10.5	10.6	9.2	10.4	10.4	10.1	8	10.1	8.9
	G	8.5	8.5	8.3	8.3	8.3	8	8	8	8	8.1

Note: Values are given volumes in cubic centimeters.
[a] M = manual, G = geometrically constrained region growth.

$cov(Z_1, Z_2) = 0.166, \theta = 0, Z_1 = 4.6, Z_2 = 3.027, V = 0.204.$
Therefore,

$$T = \frac{(Z_1 - Z_2) - (\hat{\theta})}{(\hat{V})^{\frac{1}{2}}} = 3.487.$$

The *p*-value of this test is less than 0.001, indicating the significance of the difference between the two ICCC.

We may construct a $(1 - \alpha)$ 100% confidence interval on the difference between two dependent ICCC as follows. First from the delta method we find the correlation between $\hat{\rho}_1$ and $\hat{\rho}_2$, which is given as

$$\text{Corr}(\hat{\rho}_1, \hat{\rho}_2) = \frac{\{n_1 n_2)(n_1 - 1(n_2 - 1)\}^{1/2} \rho_{12}^2}{(1 + (n_1 - 1)\rho_1)(1 + (n_2 - 1)\rho_2)}. \tag{2.8}$$

Second, the variance of the difference between the estimated ICCC is given as

$$\text{var}(\hat{\rho}_1 - \hat{\rho}_2) = \text{var}(\hat{\rho}_1) + \text{var}(\hat{\rho}_2) - 2\text{Corr}(\hat{\rho}_1, \hat{\rho}_2)\sqrt{\text{var}(\hat{\rho}_1)\text{var}(\hat{\rho}_2)},$$

where $\text{var}(\hat{\rho}_j)$ is the large sample variance of the estimated ICC given by Equation 2.5. The 95% confidence interval on $\rho_1 - \rho_2$ is thus, given by (−0.015, 0.059).

2.5 Some Results for the One-Way Random Effects Model When the Data Are Unbalanced

In this appendix we provide the interested reader with the algebra of the one-way random effects model. It has no effect on the developments in the remainder of the chapter, and may be skipped.

2.5.1 The Model

The model equation that shall be used is

$$y_{ij} = \mu + b_i + w_{ij} \tag{2.9}$$

where y_{ij} is the *j*th observation in the *i*th class, μ is a general mean, b_i is the effect on the *y*-variable of it being observed on an observational unit that is in the *i*th class, and w_{ij} is the residual error. The number of classes in the data shall be denoted by *a*, and the number of observations in the *i*th class by n_i. Thus $i = 1, 2, ..., k$ and $j = 1, 2, ..., n_i$ for $n_i \geq 1$. For balanced data there is the same number of observations in every class, say *n*, so that $n_i = n$ for every class.

The Model Equation

For the random model we must take account of b_i being a random variable. To do so, we first assume that

$$E(b_i) = 0$$

where E represents expectation over the population of b_i.

Furthermore, it is assumed that

$$\text{cov}(w_{ij}, w_{i'j'}) = 0 \text{ except for } i = i' \quad \text{and} \quad j = j'.$$

This means that the covariance between every pair of different w_{ij} terms is zero; similarly, for the b_i terms,

$$\text{cov}(b_i, b_{i'}) = 0;$$

and likewise for the covariance of each b_i with every w_{ij}:

$$\text{cov}(b_i, w_{i'j'}) = 0.$$

Now consider (4) and (5) for $i = i'$ and $j = j'$. These lead to variances, defined as

$$\text{var}(w_{ij}) = \sigma_w^2 \quad \text{and} \quad \text{var}(b_i) = \sigma_b^2. \tag{2.10}$$

These variances, σ_w^2 and σ_b^2 are called variance components because they are the components of the variance of an observation

$$\sigma_y^2 = \text{var}(y_{ij}) = \text{var}(\mu + b_i + w_{ij}) = \sigma_b^2 + \sigma_w^2. \tag{2.11}$$

Moreover, although b_i and w_{ij} are uncorrelated, the y_{ij}s are not. For those in the same subject

$$\text{cov}(y_{ij}, y_{i'j'}) = \sigma_b^2 \quad \text{for } j \neq j',$$

whereas for two different subjects

$$\text{cov}(y_{ij}, y_{i'j'}) = 0 \quad \text{for } i \neq i'.$$

2.5.2 Balanced Data

Expected Sum of Squares

As indicated in Table 2.1, the two sums of squares that are the basis of the ANOVA of balanced data from a one-way classification are

$$BSS = \sum_{i=1}^{k} n(\bar{y}_{i.} - \bar{y}_{..})^2$$

and

$$WSS = \sum_{i=1}^{k}\sum_{j=1}^{n} (\bar{y}_{ij} - \bar{y}_{..})^2.$$

The total sum of squares is

$$TSS = \sum_{i=1}^{k}\sum_{j=1}^{n} (\bar{y}_{ij} - \bar{y}_{..})^2.$$

The ANOVA method of estimation is based on deriving the expected values of SSA and SSE. One, then equates observed and expected values and solves for estimators.

With $BMS = BSS/(k-1)$ this gives

$$E(BMS) = \frac{E(BSS)}{k-1} = n\sigma_b^2 + \sigma_w^2,$$

$$E(WMS) = \frac{E(WSS)}{k(n-1)} = \sigma_w^2,$$

and

$$\hat{\rho}_1 = \frac{BMS - WMS}{BMS + (n-1)WMS}.$$

2.5.3 Normality Assumptions

The ANOVA method of estimation, although it uses sums of squares traditionally encountered in an ANOVA table, does not invoke normality. Neither is normality needed, of course, in the ANOVA table itself until F-statistics calculated from those sums of squares are used in confidence interval or hypothesis-testing context.

2.5.4 Sampling Variances of Estimators

The independence (under normality) of BSS and WSS has been established and each has a distribution that is proportional to a χ^2. From this we derive sampling variances of the estimators.

Under the normality assumption, Searle et al. (1992) showed that

$$\frac{(k-1)BMS}{n\sigma_b^e + \sigma_w^2} \sim X_{k-1}^2$$

and

$$\frac{k(n-1)WMS}{\sigma_w^2} \sim X_{k(n-1)}^2.$$

Moreover,

$$\text{var}\left(\hat{\sigma}_b^2\right) = \frac{2}{n^2}\left[\frac{\left(n\sigma_b^2 + \sigma_w^2\right)}{k-1} + \frac{\sigma_w^4}{k(n-1)}\right],$$

$$\text{var}\left(\hat{\sigma}_w^2\right) = \frac{4\sigma_w^2}{k(n-1)},$$

$$\text{cov}\left(\hat{\sigma}_b^2, \hat{\sigma}_w^2\right) = \frac{-2\sigma_w^2}{kn(n-1)}.$$

Using the delta methods, one can show, to the first order of approximation, that

$$\text{var}(\hat{\rho}_1) = \text{var}(\hat{\sigma}_b^2)\left(\frac{\partial\hat{\rho}}{\partial\hat{\sigma}_b^2}\right)^2 + \text{var}(\hat{\sigma}_w^2)\left(\frac{\partial\hat{\rho}}{\partial\hat{\sigma}_w^2}\right)^2 + 2\text{cov}(\hat{\sigma}_b^2, \hat{\sigma}_w^2)\left(\frac{\partial\hat{\rho}}{\partial\hat{\sigma}_b^2}\right)^2\left(\frac{\partial\hat{\rho}}{\partial\hat{\sigma}_w^2}\right)^2.$$

Simplifying, one gets

$$\text{var}(\hat{\rho}_1) = \frac{2(1-\rho_1)^2(1+(n-1)\rho_1)^2}{kn(n-1)}.$$

2.5.5 Unbalanced Data

The ANOVA sums of squares for unbalanced data are

$$BSS = \sum_{i=1}^{k} n_i \left(\bar{y}_{i.} - \bar{y}_{..}\right)^2 = \Sigma_i n_i \bar{y}_{i.} - N\bar{y}_{..}^2$$

and

$$WSS = \sum_{i=1}^{k} \sum_{j=1}^{n_i} \left(y_{ij} - \bar{y}_{i.}\right)^2 = \Sigma_i \Sigma_j y_{ij}^2 - \Sigma_i n_i \bar{y}_i^2$$

where $N = n_1 + n_2 + \cdots + n_k$.

The expected values of the three different terms in those expressions are now given. Derivation is, of course, based on exactly the same model as described earlier, making particular use of the results such as $E(b_i) = 0$, $E(b_i^2) = \sigma_b^2$, and $E(b_i b_i') = 0$ for $i \neq i'$; and $E(w_{ij}) = 0$.

$$E\left(\Sigma_i n_i \bar{y}_{ij}^2\right) = \Sigma_i n_j E\left(\mu + b_i + \bar{w}_{i.}\right)^2 = N\mu^2 + N\sigma_b^2 + k\sigma_w^2$$

and

$$E\left(\Sigma_i \Sigma_j \bar{y}_{ij}^2\right) = N\mu^2 + N\sigma_b^2 + N\sigma_w^2$$

$$E(BSS) = (N - \Sigma_i n_i^2/N)\sigma_b^2 + (k - 1)\sigma_w^2,$$

and

$$E(WSS) = (N - k)\sigma_w^2$$

$$BSS = \left(N - \frac{\Sigma_i n_i}{N}\right)\hat{\sigma}_b^2 + (k-1)\hat{\sigma}_w^2 \quad \text{and} \quad WSS = (N-k)\hat{\sigma}_w^2$$

for unbalanced data. Therefore,

$$\hat{\sigma}_e^2 = WMS$$

and

$$\hat{\sigma}_b^2 = \frac{(BMS - WMS)}{(N - \Sigma_i n_i^2/N)/(k-1)}$$

are the ANOVA estimators for unbalanced data. They reduce to those for balanced data.

2.5.6 Sampling Variances of Estimators

Two of the three results are easy. First because $WSS/\sigma_w^2 \sim \chi_{N-k}^2$,

$$\mathrm{var}(\hat{\sigma}_w^2) = \mathrm{var}\,(WMS) = \frac{2\sigma_w^2}{n-k},$$

$$\mathrm{cov}(\hat{\sigma}_b^2, \hat{\sigma}_w^2) = \frac{\mathrm{cov}(BMS - WMS, WMS)}{(N - \Sigma_i n_i^2/N)/(k-1)}$$

$$= \frac{2\sigma_w^2}{(N-1)(N - \Sigma_i n_i^2/N)/(k-1)},$$

$$\mathrm{var}(\hat{\sigma}_b^2) = \frac{2N}{(N^2 - \Sigma_i n_i^2)}$$

$$\times \left[\frac{N(N-1)(k-1)}{(N-k)(N^2 - \Sigma_i n_i^2)} \sigma_w^4 + 2\sigma_w^2 \sigma_b^2 \right.$$

$$\left. + \frac{N^2 \Sigma_i n_i^2 + (\Sigma_i n_i^2)^2 - 2N\Sigma_i n_i^3}{N(N - \Sigma n_i^2)} \right] \sigma_b^4.$$

Smith (1956), under the assumption of normality, showed that the asymptotic variance of ρ_1, when the data are unbalanced, is given by

$$\mathrm{var}(\hat{\rho}_1) = \frac{2(1-\rho_1)}{n_o^2} \left[\frac{\left[1 + (n_o - 1)\rho_1\right]^2}{N-k} + \frac{(1-\rho_1)\left[1 + (2n_o - 1)\rho_1\right]}{k-1} + \rho_1^2 \lambda^2 \right] \qquad (2.12)$$

where

$$\lambda = \frac{1}{(k-1)^2} \left[\sum_{i=1}^{k} n_i^2 - 2N^{-1} \sum_{i=1}^{k} n_i^3 + N^{-2} \left(\sum_{i=1}^{k} n_i^2 \right)^2 \right]$$

and $n_0 = N - \Sigma_i n_i^2/N$.

2.6 Large Sample Confidence Interval on ρ_2

As we have indicated, the corresponding model for Case 2 is a two-way random effect model. The rating score of the jth rater on the ith subject represented by the model

$$y_{ij} = \mu + b_i + r_j + w_{ij} \quad (i = 1, 2, \ldots, k;\ j = 1, 2, \ldots, n)$$

$$\text{var}(y_{ij}, y_{il}) = \sigma_B^2 \quad j \neq l;\ = 1, 2, \ldots, n. \quad \rho_2 = \frac{\sigma_b^2}{\sigma_b^2 + \sigma_r^2 + \sigma_2^2}$$

Confidence intervals on ρ_2 has been based on Satterthwaite's two-moment approximation; see, for example, McGraw and Wong (1966) and Fleiss and Shrout (1978). Recently, Cappelleri and Ting proposed a modification that provided a satisfactory performance as compared to other methods.

2.6.1 Two-Moments Approximation

Fleiss and Shrout (1978) used Satterthwaite's two-moment approximation to a linear combination of independent chi-square random variables to provide an upper and lower one-sided $100\,(1 - \alpha)\%$ confidence limits as

$$U_{FS} = \frac{F_*(BMS) - WMS}{F_*(BMS) + d_2(RMS) + d_3(WMS)},$$

$$L_{FS} = \frac{BMS - F^*(WMS)}{BMS + F^*\left[d_2(RMS) + d_3(WMS)\right]}$$

where $d_2 = n/k$, $d_3 = n - 1 - n/k$.

$F_* = F_{1-\alpha:\hat{v},k-1}$ is the upper $100\,(1 - \alpha)$ percentile of an F-distribution with degrees of freedom $\hat{v}$ in the numerator and $k - 1$ in the denominator, $F_* = F_{1-\alpha:k-1,\hat{v}}$ is the upper $100\,(1 - \alpha)$ percentile of an F-distribution with degrees of freedom $k - 1$ in the numerator and $\hat{v}$ in the denominator; and $\hat{v}$ is the approximate degrees of freedom given as

$$\hat{v} = \frac{(n-1)(k-1)\left\{n\hat{\rho}_2\left(\dfrac{RMS}{WMS}\right) + k\left[1 + (n-1)\hat{\rho}_2\right] - n\hat{\rho}_2\right\}^2}{(k-1)n^2\hat{\rho}_2\left(\dfrac{RMS}{WMS}\right)^2 + \left\{k\left[1 + (n-1)\hat{\rho}_2\right] - n\hat{\rho}_2\right\}^2}$$

An approximate two-sided $100(1 - \alpha)\%$ can be obtained by using $100(1 - \alpha/2)$ percentile of the described F-distribution.

2.6.2 Higher Moments Approximation

Seeking to improve the approximation of the one-sided lower bound confidence interval for ρ_2, Zou and McDermott (1999) developed a three- and a four-moment approximation using the Pearson system of curves. They replaced $\hat{v}$ with

$$\hat{v}_T = \frac{\left(\dfrac{\hat{a}^2\hat{u}^2}{v_1} + \dfrac{\hat{a}_2^2}{v_2}\right)^3}{\left(\dfrac{\hat{a}_1^3\hat{u}^3}{v_1} + \dfrac{\hat{a}_2^3}{v_2^2}\right)^2}$$

where

$$\hat{a}_1 = \frac{n\hat{\rho}_2}{k\left(1 - \hat{\rho}_2\right)},$$

$$\hat{a}_2 = 1 + \frac{n\left(k - 1\right)\hat{\rho}_2}{k\left(1 - \hat{\rho}_2\right)}$$

$$v_1 = n - 1, \quad v_2 = (n - 1)(k - 1) \quad \text{and} \quad \hat{u} = \frac{RMS}{WMS}.$$

2.6.3 Modified Large Sample Approach

Cappelleri and Ting (2003) discovered, through simulations, that the three- and four-moment approximation methods tend to be too conservative. They modified the three-moment approximation by adjusting the degrees of freedom of the F-distribution. Specifically, they adjusted the degrees of freedom by taking the weighted average of $\hat{v}$ and $\hat{v}_T$ such that the modified degrees of freedom is

$$\hat{v}_T = 0.1\hat{v} + 0.9\hat{v}_T.$$

This new approximation tends to provide the most accurate coverage relative to other method. The authors provided an SAS macro that may be used to construct an approximate confidence interval and $\rho_{,,}$ using the modified three-moment approximation.

2.7 Raters Agreement for Continuous Scale Measurements

Haber and Barnhart (2006) argued that the two-way ANOVA used to estimate ρ_3 requires inappropriate assumptions. These assumptions include the homogeneity of the error variances and the homogeneity of the pair-wise correlation between raters. When agreement is defined as the difference of the scores of different raters on the same subject, different index of agreement is used. This index is known as the concordance correlation coefficient (CCC). We show near the end of this section that under certain conditions the CCC coincides with the ICC. The CCC was introduced by Lin (1989), and in the case of two raters it measures agreement by assessing the variation of their linear relationship from the 45° line through the origin.

Lin (1989) introduced the CCC for two raters whose measurements on k subjects are represented by two random variables (Y_1, Y_2). Furthermore, we assume that $E(Y_i) = \mu_i$, $\text{var}(Y_i) = \sigma_i^2$, and $\text{Corr}(Y_1, Y_2) = R$.

The average difference between and pair of measurements taken on subject (i) is given by

$$E(Y_{i1} - Y_{i2})^2 = \sigma_1^2 + \sigma_2^2 + (\mu_1 - \mu_2)^2 - 2R\sigma_1\sigma_2.$$

When there is no correlation (i.e., $R = 0$), Lin (1989) defined the CCC, denoted by ρ_C as

$$\rho_C = 1 - \frac{E(Y_{i1} - Y_{i2})^2}{\sigma_1^2 + \sigma_2^2 + (\mu_1 - \mu_2)^2}$$

$$= \frac{2R\sigma_1\sigma_2}{\sigma_1^2 + \sigma_2^2 + (\mu_1 - \mu_2)^2}$$

(2.13)

making the substitution

$$\beta = \frac{\sigma_2}{\sigma_1} R$$

$$\alpha = \mu_2 - \beta\mu_1$$

$$\rho_C = \frac{2\beta\sigma_1^2}{\sigma_1^2 + \sigma_2^2 + \left[(\alpha - 0) + (\beta - 1)\mu_1\right]^2}.$$

This CCC evaluates the extent to which pairs of measurements fall on the 45° line. Any departure from this line would produce $\rho_C < 1$ even if $R = 1$.

Lin (1989) suggested a moment estimator of ρ_C given by

$$\hat{\rho}_C = \frac{2rS_1S_2}{S_1^2 + S_2^2 + (\bar{y}_1 - \bar{y}_2)}$$

where $\bar{y}_j = (1/k)\sum_{i=1}^{k} y_i$, $S_j^2 = (1/k)\sum_{i=1}^{k}(y_{i_j} - \bar{y}_j)^2$, and $r = \sum_{i=1}^{k}(y_{i1} - \bar{y}_1)$ $(y_{i2} - \bar{y}_2)/S_1S_2$ are the sample moment estimators of Pearson correlation coefficient.

Let us assume that $(y_{11}, y_{12}), (y_{21}, y_{22}), \ldots, (y_{k1}, y_{k2})$ is a random sample from a bivariate normal distribution with mean vector $\mu = (\mu_1 \quad \mu_2)^T$ and variance–covariance matrix

$$\Omega = \begin{pmatrix} \sigma_1^2 & \sigma_{12} \\ \sigma_{21} & \sigma_2^2 \end{pmatrix}$$

where $\sigma_{12} = R\sigma_1\sigma_2$.

Under fairly general conditions, Lin (1989) showed that $\hat{\rho}_C$ is asymptotically normally distributed with mean ρ_C and variance

$$\text{var}(\hat{\rho}_C) = \frac{1}{k-2}\left[\frac{(1-R^2)\rho_C^2(1-\rho_C^2)}{R^2} + \frac{4\rho_C^3(1-\rho_C)u^2}{R} - 2\rho_C^4u^4/R^2\right] \quad (2.14)$$

where $u = (\mu_1 - \mu_2)/(\sqrt{\sigma_1\sigma_2})$.

To improve on the normal approximation, Lin (1989) employed Fisher's (1925) transformation

$$Z(\hat{\rho}_C) = \frac{1}{2}\ln\left[\frac{1+\hat{\rho}_C}{1-\hat{\rho}_C}\right].$$

For large sample he argued that

$$Z(\hat{\rho}_C) \sim N(Z(\rho_C), \sigma_Z^2)$$

where

$$\sigma_Z^2 = \frac{1}{k-2}\left[\frac{(1-R^2)\rho_C^2}{(1-\rho_C^2)R^2} + \frac{4\rho_C^3(1-\rho_C)u^2}{R(1-\rho_C^2)^2} - \frac{2\rho_C^4u^2}{R^2(1-\rho_C^2)^2}\right]. \quad (2.15)$$

If there are more than two raters ($n > 2$) and when the assumption of homogeneity of error variance and the homogeneity of the pair-wise correlations between raters are not satisfied, Carrasco and Jover (2003) showed the ρ_C can be estimated by

$$\hat{\rho}_{CC} = \frac{2\sum_{i=1}^{n-1}\sum_{j=1}^{n} R_{ij} \cdot S_i S_j}{(n-1)\sum_{j=1}^{n} S_j^2 + \sum_{i=1}^{n-1}\sum_{j=1}^{n}(\bar{Y}_i - \bar{Y}_j)^2} \quad n > 2.$$

Here, R_{ij} is the pair-wise correlation between raters i and j. Note that when $n = 2$, $\hat{\rho}_{CC}$ reduces to $\hat{\rho}_C$.

2.8 Estimating Agreement When Gold Standard Is Present

In previous sections of this chapter we focused on measurement reliability and agreement between two or more rating methods. Here are however, situations when one of the raters is error free and may be considered as the gold standard. St. Laurent (1998) provided several examples. Wax et al. (1992) compared two methods used to assess blood alcohol concentration (BAC). The first method is the rapid electrochemical meter, and the gold standard BAC determined by blood immunoassay.

The methodological approach to assess agreement between faulty methods and gold standard does not substantially differ from the traditional approach. The fundamental difference is that the gold standard measurements are observable. Consider the model

$$Y_i = T_i + e_i,$$

where Y_i is the faulty method measurement on the ith subject ($i = 1, 2, \ldots, k$); T_i is the corresponding gold standard method. Here T_i represents an observable random variable with mean μ and variance σ_T^2. The e_i is a measurement error independent of T_i, with mean 0 and variance σ_e^2. This model indicates that for each subject the faulty method would provide perfect agreement with gold standard if there is not additive error measurement. Under these conditions, $\text{cov}(Y_i, T_i) = \sigma_T^2$ and $\text{var}(Y_i) = \sigma_T^2 + \sigma_e^2$, and it follows that

$$\rho_g = \frac{\sigma_T^2}{\sigma_T^2 + \sigma_e^2}.$$

This is again the intraclass correlation.

St. Laurent (1998) suggested as an appropriate estimator for ρ_g—a quantity that depends on $S_T^2 = \sum_{i=1}^{k}(T_i - \bar{T})^2$ and S_D^2, where $T = (1/k)\sum_{i=1}^{k} T_i$, $S_D^2 = \sum_{i=1}^{k}(Y_i - T_i)^2$,

$$\hat{\rho}_g = \frac{S_T^2}{S_T^2 + S_D^2}.$$

St. Laurent developed an inferential procedure on ρ_g as follows: If e_i and T_i are normally distributed, then the statistic

$$\frac{k-1}{k}\left(\frac{1-\hat{\rho}_g}{\hat{\rho}_g}\right)$$

is distributed as $(1 - \rho_g/\rho_g)F_{k,k-1}$, where $F_{k,k-1}$ is an F random variable with k and $k-1$ degrees of freedom. He then constructed a one-side $100(1-\zeta)\%$ lower confidence interval on ρ_g as

$$\frac{F_L}{F_L + \dfrac{k-1}{k}\left(\dfrac{1-\hat{\rho}_g}{\hat{\rho}_g}\right)} < \rho_g < 1$$

where F_L is such that $\Pr\left[F_{k,k-1} < F_L\right] = \zeta$.

2.9 Several Faulty Methods and a Gold Standard

In many applications, there are several faulty methods of measurements that need to be compared to the gold standard. To illustrate the situation we assume that we have two faulty methods and the measurements:

$$Y_{i1} = T_i + e_{i1}$$

$$Y_{i2} = T_i + e_{i2}$$

Again, we assume that $E(T_i) = \mu$, $E(e_{ij}) = 0$, $\text{var}(T_0) = \sigma_T^2$. Furthermore, we assume that $\text{var}(e_{ij}) = \sigma_j^2$, and $\text{cov}(e_{i1}, e_{i2}) = \sigma_{12}$. A consequence of the above assumptions is that

$$\text{cov}(Y_{i1}, Y_{i2}) = \sigma_T^2 + \sigma_{12}.$$

Therefore,

$$\rho_{g(j)} = \frac{\sigma_T^2}{\sigma_T^2 + \sigma_j^2} \quad j = 1, 2.$$

As noted by St. Laurent (1998), $\rho_{g(j)}$ ($j = 1, 2$) do not depend on the covariance parameter σ_{12}.

Denote $S_{D(j)}^2 = \sum_{i=1}^{k}(Y_{ij} - T_i)^2$, therefore a natural estimator for $\rho_{g(j)}$ is

$$\hat{\rho}_{g(j)} = \frac{S_T^2}{S_T^2 + S_{D(j)}^2}.$$

A natural question to ask is whether one is in closer agreement with the gold standard than the other. That is, whether one is interested in testing the null hypothesis $H_0 : \rho_{g(1)} = \rho_{g(2)}$ versus $H_1 : \rho_{g(1)} \neq \rho_{g(2)}$. Since $\rho_{g(1)} = \rho_{g(2)}$, if and only if $\sigma_1^2 = \sigma_2^2$, the above hypotheses are equivalent to testing $H_0 : \sigma_1^2 = \sigma_2^2$ versus $H_1 : \sigma_1^2 \neq \sigma_2^2$. Now, for fixed j, $D_{ij} = Y_{ij} - T_i$ are independent with $E(D_{ij}) = 0$, $\text{var}(D_{ij}) = \sigma_j^2$ and $\text{cov}(D_{i1}, D_{i2}) = \sigma_{12}$. Therefore, the problem of testing equality of agreement is equivalent to testing the equality of error variances. This problem has been examined by many authors, Pitman (1939), Morgan (1939), Grubbs (1948), Maloney and Rastogi (1970), and Jaech (1985). Then Shukla (1973) showed that, if we define

$$W_i = D_{i1} + D_{i2}$$

$$U_i = D_{i1} - D_{i2}$$

then

$$\text{cov}(W_i, U_i) = \sigma_1^2 - \sigma_2^2$$

$$= \rho_{wu}\sigma_w\sigma_u$$

where ρ_{wu} is the Pearson's correlation between W_i and U_i. Let $\hat{\rho}_{wu}$ denote the estimate of ρ_{wu}. Hence, testing the equality of σ_1^2 and σ_2^2 is equivalent to testing that the correlation between W_i and U_i is zero. Therefore, $H_0 : \sigma_1^2 = \sigma_2^2$ is rejected whenever

$$\tau = \hat{\rho}_{wu}\sqrt{\frac{k-2}{1-\hat{\rho}_{wu}}}$$

exceeds $|\tau_{\alpha/2}|$ where $\tau_{\alpha/2}$ is the cut-off point found in the t-table at $100(1 - \alpha/2)\%$ coefficient, and $(k - 2)$ degrees of freedom.

2.10 Sample Size Requirements for Design of Reliability Study under One-Way ANOVA

2.10.1 Introduction

In some reliability studies investigators report the average of m observations by the jth rater for the ith subject. In this case, the ICC takes the form

$$\rho_m = \frac{\sigma_b^2}{\sigma_b^2 + \sigma_e^2/m} = \frac{m\rho_1}{1+(m-1)\rho_1}.$$

Clearly, ρ_m is a monotonic increasing function of m. However, if $\rho_1 = 1$, then $\rho_m = 1$ and increasing m will not increase the reliability. Same remarks hold for $\rho = 0$. Also, one might want to determine the number of replications m for which ρ_m exceeds a certain level of ρ^* (say). In this case

$$\rho_m \geq \rho^* \Rightarrow \frac{m\rho_1}{1+(m-1)\rho_1} \geq \rho^*.$$

Solving the above inequality for m yields

$$m \geq \frac{\rho^*\left(1-\rho_1\right)}{\rho_1\left(1-\rho^*\right)}$$

(see Kraemer, 1979).

As discussed in the previous sections, measurement errors can seriously affect statistical analysis and interpretation; it therefore becomes important to assess the amount of such errors by calculation of a reliability coefficient or a coefficient of agreement if the assessments are binary.

Although the topic of reliability has gained much attention in the literature, investigations into sample size requirements remain scarce. In this chapter, we discuss the issue of sample size requirements to conduct a reliability study for both continuous and binary assessments.

2.10.2 Case of Continuous One-Way ANOVA

2.10.2.1 Power Considerations

We assume, as in Chapter 2, a one-way random effects model, which is frequently used to investigate reliability

$$y_{ij} = \mu + s_i + e_{ij}$$

where μ is the grand mean of all measurements in the population, s_i reflects the effect of the characteristic under measure for subject i, e_{ij} is the error of measurement, $j = 1, 2, \ldots, n$, $i = 1, 2, \ldots, k$.

Suppose we assume further that the subject effects $\{s_i\}$ are normally and identically distributed with mean zero and variance σ_s^2, the errors $\{e_{ij}\}$ are normally and identically distributed with mean zero and variance σ_e^2, and the $\{s_i\}$ and $\{e_{ij}\}$ are independent. Then the population ICC is $\rho = \sigma_s^2/(\sigma_s^2 + \sigma_e^2)$. The sample intraclass correlation

$$\hat{\rho} = \frac{MSB - MSW}{MSB + (n-1)MSW} = \frac{F-1}{F+n-1}$$

estimates ρ.

Donner and Eliasziw (1987) discussed statistical power consideration to estimate values of k and n required to test $H_1 : \rho = \rho_0$ versus $H_1 : \rho > \rho_0$, where ρ_0 is a specified criterion value of ρ.

For the case $n = 2$ (i.e., test–retest data), and to establish asymptotic properties for $\hat{\rho}$, we may use Fisher's (1958) normalizing transformation for $\hat{\rho}$, which is analogous to the well-known Fisher transformation of the Pearson product–moment (or interclass correlation). He showed that $u = (1/2)\ln((1+\hat{\rho})/(1-\rho))$ is very nearly normally distributed with mean $u(\rho) = (1/2)\ln((1+\rho)/(1-\rho))$ and variance $\sigma_u^2 = (k - 3/2)^{-1}$.

Note that from Chapter 2,

$$\hat{\rho} = \frac{MSB - MSW}{MSB + MSW}$$

is the ANOVA estimator for ρ when $n = 2$.

Let z_α and z_β denote the values of the standard normal distribution corresponding to the chosen level of significance α and power $(1 - \beta)$.

The required number of subjects for testing $H_0 : \rho = \rho_0$ versus $H_1 : \rho = \rho_1 > \rho_0$ is obtained directly from the above theory as

$$k = \left[\frac{(z_\alpha + z_\beta)}{\mu(\rho_0) - \mu(\rho_1)} \right]^2 + \frac{3}{2}.$$

Table 2.7 gives the required values of k according to the values of ρ_0 and ρ_1, $\alpha = 0.05$ and $\beta = 0.20$. Note that $z_{0.05} = 1.64$ and $z_{0.2} = 0.84$.

The results in Table 2.7 indicate that the required sample size k depends critically on the values of ρ_0 and ρ_1, and on their difference in particular. So, for example, much more effort is required to distinguish ρ values that differ by 0.1 compared to those with a difference of 0.2. Also the larger samples are

TABLE 2.7

Number of Subjects k for $\alpha = 0.05$ and $\beta = 0.20$

ρ_0	ρ_1	k
0.2	0.6	27
0.2	0.8	9
0.4	0.6	86
0.8	0.9	46
0.6	0.8	39

required in association with relatively small values of ρ, for a given difference $\rho_1 - \rho_0$.

For n and k unknown, Walter et al. (1998) developed a simple approximation that allows the calculation of required sample size for the number of subjects k, when the number of replicates n is fixed. Their approximation uses a single formula, and avoids the intensive numerical work needed with the exact methods as in Donner and Eliasziw (1987). Furthermore, it permits the investigator to explore design options for various parameter values. The interest is in testing $H_0 : \rho = \rho_0$ versus $H_1 : \rho = \rho_1$. The hypothesis H_0 is tested using $MSB/MSW = (1 + (n-1))\hat{\rho}/(1 - \hat{\rho})$ from the ANOVA, and where $\hat{\rho} = (MSB - MSW)/(MSB + (n-1)MSW)$ is the sample estimator of ρ. The critical value for the test statistic is CF_{α,v_1,v_2}, where $C = 1 + [n\rho_0/(1 - \rho_0)]$ and F_{α,v_1,v_2} is the $100(1-\alpha)$ percent point in the cumulative F-distribution with (v_1, v_2) degrees of freedom, where $v_1 = k - 1$ and $v_2 = k(n-1)$.

As described by Donner and Eliasziw (1987), at $\rho = \rho_1$, the test H_0 has power

$$1 - \beta = \Pr\left[F \geq C_0 F_{\alpha,v_1,v_2}\right]$$

where β is the type II error and $C_0 = (1 - n\phi_0)/(1 + n\phi)$, with $\phi_0 = \rho_0/(1 - \rho_0)$, and $\phi = \rho_1/(1 - \rho_1)$.

To solve the power equation, Walter et al. (1998) used a result by Fisher (1925), regarding the asymptotic distribution of $Z = (1/2)\log F$. Omitting the details, the estimated number of subjects is

$$k = 1 + \frac{2n(z_\alpha + z_\beta)^2}{(n-1)(\ln C_0)^2} = 1 + \frac{nA(\alpha,\beta)}{(n-1)(\ln C_0)^2}$$

where z_α is the $100(1-\alpha)$ per cent in the cumulative unit normal distribution, and $A(\alpha,\beta) = 2(z_\alpha + z_\beta)^2$. Table 2.8 gives the values of $A(\alpha,\beta)$ for combinations of α and β.

Table 2.9 shows the required values of k for typical values of n, and according to the values of ρ_0 and ρ_1, with $\alpha = 0.05$ and $\beta = 0.10$.

TADLE 2.0

$A(\alpha, \beta) = 2(z_\alpha + z_\beta)^2$

		$1 - \beta$	
α	0.80	0.90	0.95
0.10	8.99	13.10	17.05
0.05	12.30	17.05	21.52
0.01	20.10	26.06	31.52

The results in Table 2.9 indicate that the required sample size k depends on the values of ρ_0 and ρ_1 and on their difference in particular.

2.10.2.2 Fixed Length Requirements of Confidence Interval

Recently, Giraudeau and Mary (2001) (GM) and Bonett (2002) argued that the approach of hypothesis testing may not be appropriate while planning a reproducibility study. This is because one has to specify both the values of ρ_0 and ρ_1, which may be difficult to choose and questionable. A conceivable way to plan the study is therefore to focus on the width of the confidence interval (CI). Indeed, when there is a unique sample and no comparative purpose, the results of a reproducibility study are usually expressed as a point estimate of ρ and its associated confidence interval. The sample size calculations are then aimed at achieving sufficient precision of the estimate.

The approximate width of a 95% confidence interval on ρ is equal to $2z_{\alpha/2}(\mathrm{var}(\hat{\rho}))^{1/2}$ where

$$\mathrm{var}(\hat{\rho}) = \frac{2(1-\rho)^2 (1 + (n-1)\rho)^2}{kn(n-1)} \tag{2.16}$$

is the approximate variance of the intraclass correlation estimator ρ and $z_{\alpha/2}$ is the point on a standard normal distribution exceeded with probability

TABLE 2.9

Approximate Sample Size ($\alpha - 0.05$, $\beta - 0.10$)

ρ_0	ρ_1	n	$\ln C_0$	k
0.2	0.4	10	−0.784	32
0.4	0.8	20	−1.732	6
0.6	0.8	10	−0.941	22
0.8	0.9	10	−0.797	31
0.2	0.6	2	−/981	36

$\alpha/2$. It is known that Equation 2.16 is accurate when $k \geq 30$. An approximation to the sample size that will yield an exact confidence interval for ρ having desired width w is obtained by setting $w = 2z_{\alpha/2}(\text{var}(\rho))^{1/2}$, replacing ρ with a planning value $\bar{\rho}$, and then solving for k to give

$$k = 8z_{\alpha/2}^2 (1 - \bar{\rho})^2 (1 + (n-1)\bar{\rho})^2 / \{w^2 n(n-1)\}, \qquad (2.17)$$

which should be rounded up to the nearest integer.

The approximation suggested by Bonett (2002) is $k = k + 1$ where k is given by Equation 2.17.

Table 2.10 gives the required sample size for planned values of $\rho = 0.6, 0.7, 0.8$, and 0.9 (the most practically assigned values of ρ in a typical reliability study), $w = 0.2$, $\alpha = 0.05$, and $n = 2, 3, 5, 10$.

As can be seen from Table 2.10, the sample size requirement is a decreasing function of n for any given value of $\bar{\rho}$. Thus, it may be less costly to increase the number of measurements per subject than to increase the number of subjects. The advantages of interval estimation over hypothesis testing have been discussed by many others, but an additional issue should be considered. A planned value of ρ_1 is needed for sample size determination in both hypothesis testing and interval estimation. Bonett (2002) argued that the effect of an inaccurate planning value is more serious in hypothesis testing applications. For example, to test $H_0 : \rho = 0.7$ at $\alpha = \beta = 0.05$, with $n = 3$, the required sample size by Walter et al. (1998) is about 376, 786, and 167 for $\rho_1 = 0.725, 0.75$, and 0.80, respectively. In comparison, the sample size required to estimate ρ with a 95 per cent confidence interval width of 0.2 is 60, 52, and 37 for $\bar{\rho} = 0.725, 0.75$, and 0.80, respectively.

2.10.2.3 Efficiency Requirements

Given that reliability studies are estimation procedures, it is natural to base the sample size calculations on the attainment of a specified level of precision in the estimate of ρ. In this section, it is assumed that the investigator is interested in the number of replicates n, per subject, so that the variance of

TABLE 2.10

Values of k^* for Planned Values of ρ, and $w = 0.2$

$\bar{\rho}$	2	3	5	10
		n		
0.6	158	100	71	57
0.7	101	67	51	42
0.8	52	36	28	24
0.9	15	11	9	8

the estimator for ρ is minimized given that the total number of measurements is, due to cost limitations constrained to be $N = nk$ *a priori.*

Substituting $N = nk$ in Equation 2.16 gives

$$\text{var}(\bar{\rho}) = f(n,\rho) = \frac{2(1-\rho)^2\left(1+(n-1)\rho\right)^2}{N(n-1)}.$$

A necessary condition for $f(n, \rho)$ to have a minimum is that $(\partial f/\partial n) = 0$, and the sufficient condition is that $(\partial^2 f/\partial n^2) > 0$ (see Rao, 1984, p. 53).

Differentiating $f(n, \rho)$ with respect to n, equating to zero, and solving for n, we get

$$n_0 = \frac{(1+\rho)}{\rho}, \tag{2.17}$$

Moreover,

$$\left.\frac{\partial^2 f}{\partial n^2}\right| = \frac{\left(4\rho^3(1-\rho)^2\right)}{N} > 0$$

$$n = n_0$$

and the sufficient condition for a unique minimum is therefore satisfied. Note that the range of ρ is strictly positive, since within the framework of reliability studies, negative values of ρ are meaningless.

Equation 2.17 indicates that, when $\rho = 1$, then $n_0 = 2$ is the minimum number of replicates needed from each subject. The smaller the value of ρ, the larger is the number of replicates, and hence, smaller number of subjects would be recruited. Table 2.11 illustrates the optimal combinations (n, k) that minimize the variance of $\hat{\rho}$ for different values of ρ.

TABLE 2.11

Optimal Combinations of (k, n) for Which the Variance of r Is Minimized

		\rho								
		0.1	0.2	0.3	0.4	0.5	0.6	0.7	0.8	0.9
	N	11	6	4.3	3.5	3	2.7	2.4	2.25	2.1
60	k	4.45	10	13.8	17.1	20	22.5	24.7	26.7	28.4
	var(r)	(0.011)	(0.017)	(0.020)	(0.019)	(0.017)	(0.013)	(0.008)	(0.004)	(0.001)
N 90	k	8.18	15	20.8	25.7	30	33.75	37	40	42.6
	var(r)	(0.007)	(0.011)	(0.013)	(0.013)	(0.011)	(0.008)	(0.006)	(0.003)	(0.001)
120	k	10.9	20	27.7	34.3	40	45	49.4	53.3	56.8
	var(r)	(0.005)	(0.008)	(0.010)	(0.010)	(0.008)	(0.006)	(0.004)	(0.002)	(0.001)

From Table 2.11, we observe that

1. Because $N = nk$ is a fixed *a priori*, higher number of replicates (n) would lead to a much smaller number of subjects and hence, there is a loss in precision of the estimated reliability. In other words, large n means that a smaller number of subjects would be recruited—that is, a less representative sample of the population of subjects being investigated.

2. When ρ is expected to be larger than 0.6, which is the case in many reliability studies, it is recommended that the study be planned with not more than two or three replicates per subject.

3. The above guidelines are indeed quite similar to those made by GM who based their sample size calculations on the achievement of a specific width for the 95% confidence interval. This is also consistent with the results reported in Table 2.3 of Walter et al. (1998).

2.10.3 Nonnormal Case

As indicated above, the sampling distribution and formula for the variance of the reliability estimates rely on the normality assumptions, despite the fact that real data seldom satisfy these assumptions. We may expect that normality would be only approximately satisfied at best. A similar problem exists for statistical inference in the one-way random effect model ANOVA, though it has been found that the distribution of the ratio of mean squares is quite robust with respect to nonnormality under certain conditions. Scheffé (1959) investigated the effects of nonnormality, and concluded that it has little effect on inference on mean values but serious effects on inferences about variances of random effects whose kurtosis γ differs from zero (p. 345). He also noted that "the direction of the effects is such that the true α level of a $100(1 - \alpha)\%$ confidence interval will be greater than the nominal α if $\gamma_s > 0$, and the magnitude of the effect increases with the magnitude of γ_s, where γ_s corresponds to the kurtosis of s_i in Equation 2.18. Although his conclusions were based on the inference of the variance ratio $\phi = \sigma_s^2/\sigma_e^2$, they may have similar implications for the reliability parameter $\rho = \phi/(1 + \phi)$.

Tukey (1956) obtained the variance of the variance component estimates under various ANOVA models by employing "polykeys." For the one-way random affects model, together with the delta method (Kendall and Ord, 1989) it can be shown that to a first-order approximation,

$$\text{var}(\hat{\rho}) = \frac{2(1 - \rho)^2 (1 + (n - 1)\rho)^2}{kn(n - 1)} + \rho^2 (1 - \rho)^2 \left[\frac{\gamma_s}{k} + \frac{\gamma_e}{kn} \right] \tag{2.18}$$

where $\gamma_s = E(s_i^4)/\sigma_s^4$ and $\gamma_e = E(e_{ij}^4)/\sigma_e^4$ (Hemmersley, 1949; Singhal, 1984).

Note that when $\gamma_s = \gamma_e = 0$, $\mathrm{var}(\hat{\rho})$ reduces to $\mathrm{var}(\hat{\rho})$ for the normal case. Following the same optimization procedure as in Section 2.9.2.3, we find that the optimal value for n, say n^*, is

$$n^* = 1 + \frac{1}{\rho\sqrt{1 + \gamma_s}}.$$

Remarks

1. Clearly, when $\gamma_s = 0$, then $n^* = 0$. Moreover, for large values of γ_s (increased departure from normality) a smaller number of replicates are needed, implying that a proportionally larger number of subjects (k) should be recruited to ensure precise estimate of ρ. We therefore, recommend the same recruiting strategy as in the normal case.

2. Note that the error distribution does not affect the number of replicates, however, both the error distribution and the between-subjects random effect distributions affect the precision of $\hat{\rho}$. But as can be seen from Equation 2.18, if $N = nk$ is large, then the influence of γ_e on the estimated precision is much smaller than the influence of γ_s.

2.10.4 Cost Implications

It has long been recognized that funding constraints determine the recruiting cost of subjects needs for a reliability study. Choice of a small sample will lead to a study that may produce imprecise estimate of the reliability coefficient. On the other hand, too large a sample may result in waste of both time and money. The crucial decision in a typical reliability study is to balance the cost of recruiting subjects with the need for a precise estimate of ρ. There have been attempts to address the issue of power, rather than precision, in the presence of funding constraints. Eliasziw and Donner (1987) presented a method to determine the number of subjects, k, and the number of replications, n, that minimize the overall cost of conducting a reliability study, while providing acceptable power for tests of hypotheses concerning ρ. They also provided tables showing optimal choices of k and n under various cost constraints.

In this section, we shall estimate the combinations (n, k) that minimize the variance of $\hat{\rho}$, as given by Equation 2.16, subject to cost constraints. In our attempt to construct a flexible cost function, we adhere to the general guidelines identified by Flynn et al. (2002) and Eliasziw and Donner (1987). First, one has to identify approximately the sampling costs and overhead costs. The sampling cost depends primarily on the size of the sample, and includes data collection costs, travel costs, management, and other staff cost. On the other hand, overhead costs remain fixed regardless of sample size, such as the cost of setting the data collection form. Following Sukhatme et al. (1984, p. 284), we assume that the overall cost function is given as

$$C = c_0 + kc_1 + nkc_2$$

where c_0 is the fixed cost, c_1 the cost of recruiting a subject, and c_2 is the cost of conducting one observation.

Using the method of Lagrange multipliers (Rao, 1984), we form the objective function G as

$$G = V(k,n,\rho) + \lambda(C - c_0 - kc_1 - nkc_2) \tag{2.19}$$

where

$$\mathrm{var}(\hat{\rho}) = V(k,n,\rho) = \frac{2(1-\rho)^2\left(1+(n-1)\rho\right)^2}{kn(n-1)}$$

and λ is the Lagrange multiplier.

The necessary conditions for the minimization of G are $\partial G/\partial n = 0$, $\partial G/\partial k = 0$, and $\partial G/\partial \lambda = 0$, and the sufficient conditions for $\mathrm{var}(\hat{\rho}) = V(k,n,\rho)$ to have a constrained relative minimum are given by a theorem in Rao (1984, p. 68). Differentiating with respect to n, k, and λ, and equating to zero we obtain

$$n^3\rho c_2 - n^2 c_2(1-\rho) - nc_1(2-\rho) + (1-\rho)c_1 = 0 \tag{2.20}$$

$$\lambda = \frac{2(1-\rho)^2\left(1+(n-1)\rho\right)\left(1-2n+(n-1)\rho\right)}{k^2 n^2 (n-1)^2 c_2}$$

and

$$k = \frac{C - c_0}{c_1 + nc_2}. \tag{2.20a}$$

The third degree polynomial Equation 2.20 has three roots. Using Descartes' rule of signs, we predict that there are two positive or two complex conjugate roots and exactly one negative root. Furthermore, since $c_1, c_2 > 0$ and $0 < \rho < 1$ we conclude that there are indeed two (real) positive roots, one of which is always between 0 and 1. This conveniently leaves us with only one relevant optimal solution for n. An explicit expression for this optimal solution, which is obtained using the "solve" function of the Symbolic Toolbox of the MATLAB® Software (The MathWorks, Inc., Natick, MA), is given in Equation 2.21.

The optimal solution for n, that is, the relevant root of Equation 2.10 is

$$n_{opt} = \frac{1}{3}\left(\frac{\sqrt[3]{A}}{\rho} - B + \frac{1+\rho}{\rho} \right) \qquad (2.21)$$

where

$$A = 9R\left(\rho^3 - \rho^2 + \rho\right) + \left(\rho+1\right)^3 + 3\rho$$

$$\times \sqrt{3R\left((R+1)^2 \rho^4 - \left(6R^2 + 4R - 2\right)\rho^3 + 12R(R+1)\rho^2 - \left(8R^2 + 10R + 2\right)\rho - R - 1\right)}$$

$$B = \frac{3R\rho(\rho - 2) - (p+1)^2}{\rho A^{1/3}}$$

and

$$R = \frac{c_1}{c_2}.$$

Once the value of n is determined, for given $C - c_0$, c_1, c_2 and ρ, we substitute it in Equation 2.20a to determine the corresponding optimal k.

The results of the optimization procedure appear in Table 2.12 for ρ equal to 0.7, 0.8, 0.9, and c_1 and c_2 equal to 0.25, 0.5, 1, 3, 15, 25. Without loss of generality, we set $C - c_0 = 100$.

It is apparent from Table 2.12 that when c_1 increases (the cost per subject), the required number of subjects k decreases, while the number of measurements per subject (n) increases. However, when c_2, the total variable cost increases, both k and n decrease. On the other hand, when c_1 and c_2 are fixed, an increase in the reliability coefficient ρ, would result in a decrease in the number of replicates and an increase in the number of subjects. This trend reflects two intuitive facts: the first is that it is sensible to decrease the number of items associated with a higher cost, and increase those with lower cost; the second is that when ρ is large (high reproducibility) then fewer number of replicates per subject are needed, while higher number of subjects should be recruited to ensure that ρ is estimated with appreciable precision. This remark is similar to the conclusion reached in the previous section, when costs were not explicitly considered. We note that at the higher levels of c_1 and c_2, the optimal allocation is quite stable with respect to changes in sampling cost. This is desirable since it is often difficult to forecast the exact cost

TABLE 2.12

Optimal Values of k and n that Minimize $var(\hat{\rho})$ for $\rho - 0.7, 0.8, 0.9$ and Different Values of c_1 and c_2 and $C - c_2 - 100$

c2	ρ	0.25 n	k	0.5 n	k	1 n	k	3 n	k	5 n	k	15 n	k	25 n	k
	0.7	3	100	3.4	73	4.1	49	6	22	7.33	15	11.73	6	14.8	4
0.25	0.8	2.76	106	3.15	78	3.77	51	5.45	23	6.65	15	10.60	6	13.35	4
	0.9	2.57	112	2.9	81	3.48	53	5	24	6.07	15	9.64	6	12.12	4
	0.7	2.74	62	3	50	3.44	37	4.69	19	5.6	13	8.68	5	10.82	3
0.5	0.8	2.52	66	2.76	53	3.15	39	4.27	19	5.10	13	7.86	5	9.78	4
	0.9	2.36	70	2.5	56	2.92	41	3.93	20	4.67	14	7.16	5	8.90	4
	0.7	260	35	2.74	31	3	25	3.8	15	4.42	11	6.54	5	8.03	3
1	0.8	2.40	38	2.53	33	2.76	26	3.48	15	4.03	11	5.93	5	7.28	3
	0.9	2.24	40	2.36	35	2.57	28	3.21	16	3.71	11	5.43	5	6.65	3
	0.7	2.5	13	2.54	12	2.64	11	3	8	3.30	7	4.42	4	5.25	2
3	0.8	2.3	14	2.35	13	2.44	12	2.76	9	3.0	7	4.03	4	4.77	3
	0.9	2.1	15	2.20	14	2.28	13	2.57	9	2.81	7	3.72	4	4.38	3
	0.7	2.5	8	2.5	8	2.56	7	2.79	6	3.00	5	3.80	3	4.42	2
5	0.8	2.3	8	2.3	8	2.37	8	2.58	6	2.76	5	3.48	3	4.03	2
	0.9	2.1	9	2.2	9	2.22	8	2.4	7	2.57	6	3.21	3	3.72	2
	0.7	2.4	3	2.4	3	2.47	3	2.56	2	2.64	2	3	2	3.30	2
15	0.8	2.3	3	2.3	3	2.29	3	2.37	3	2.44	2	2.76	2	3.03	2
	0.9	2.1	3	2.1	3	2.15	3	2.22	3	2.28	3	2.57	2	2.81	2
	0.7	2.4	2	2.4	2	2.45	2	2.51	2	2.56	2	2.8	2	3.00	1
25	0.8	2.3	2	2.3	2	2.27	2	2.32	2	2.37	2	2.58	2	2.76	1
	0.9	2.1	2	2.1	2	21	2	2.17	2	2.22	2	2.4	2	2.57	1

prior to the initiation of the study. Finally, we also note that by setting $c_1 = 0$ and $c_2 = 1$ in Equation 2.21, we get $n_{opt} = (1 + \rho)/\rho$ as in Equation 2.17. This means that a special cost structure is implied in the optimal allocation. Moreover, setting $\rho = 1$ in Equation 2.21, gives $n_{opt} = 1 + (1 + c_1/c_2)^{1/2} \geq 2$, emphasizing that the ratio c_1/c_2 is an important factor in determining the optimal allocation of (n, k).

Example 2.3

To assess the accuracy of Doppler echocardiography (DE) in determining aortic valve area (AVA) prospective evaluation on patients with aortic stenosis (AS), an investigator wishes to demonstrate a high degree of reliability ($\rho = 90\%$) of estimating AVA using the "velocity integral method." The main interest is in determining the optimal number of patients (k) and the number of readings per patient, needed to provide a highly accurate estimate of ρ, subject to cost constraints. Suppose that the total cost of making the study is held at $1600. We assume that the cost (e.g., fuel and preparation costs) of traveling of a patient from the health center to the tertiary hospital, where the procedure is done, is $15. The administrative cost of the procedure and the cost of using the DE is $15 per visit. It is assumed that c_0, the overhead cost is absorbed by the hospital. From Table 2.12, the optimal allocation for n is 3. From Equation 2.20a, the number of subjects are

$$k = \frac{1600}{15 + 3 \times 15} = 27,$$

that is, we need 27 patients, with three measurements each. The minimized value of var($\hat{\rho}$) = $V(k,n,\rho)$ is 0.001.

2.11 Sample Size for Case 2

In many reliability studies several observers (raters) agree that the two-way model may be appropriate. When we consider raters as a random sample from a specific population, Shrout and Fleiss recommended using the two-way random effects model (Case 2). Recall that for the Case 2 model we have n raters each evaluate the same group of k subjects. Raters and subjects are considered random. The score assigned by the jth rater to the ith subject is y_{ij} so that

$$y_{ij} = \mu + b_i + r_j + w_{ij}.$$

Here μ is the population mean, b_i is the effect of subjects assumed to be normally distributed with mean 0 and variance σ_b^2, r_j represents the effect of the jth rater assumed to be normally distributed with mean 0 and variance σ_r^2. The error term w_{ij} is also assumed normally distributed with mean 0 and variance σ^2. The ICC in this case is

$$\rho_2 = \frac{\sigma_b^2}{\sigma_b^2 + \sigma_r^2 + \sigma_2^2}.$$

The corrected mean squares for the two-way ANOVA are

$$BMS = n \sum_{i=1}^{k} (\bar{y}_{i.} - \bar{y}_{..})^2 \big| (k-1)$$

$$RMS = k \sum_{j=1}^{n} (\bar{y}_{.j} - \bar{y}_{..})^2 \big| (n-1)$$

$$MSE = \sum_{j=1}^{n}\sum_{i=1}^{k}\left(y_{ij} - \bar{y}_{i.} - \bar{y}_{.j} + \bar{y}_{..}\right)^2 \Big|(n-1)(k-1)$$

where

$$\bar{y}_{..} = \frac{1}{nk}\sum_{i=1}^{k}\sum_{j=1}^{n} y_{ij}$$

is the ground mean.

The ANOVA estimator of ρ_2 is obtained on equating the above mean-squares to their expectations.

$$E(BMS) = n\sigma_b^2 + \sigma_w^2$$

$$E(RMS) = k\sigma_r^2 + \sigma_w^2$$

$$E(MSE) = \sigma_w^2.$$

Therefore,

$$\hat{\sigma}_w^2 = MSE$$

$$\hat{\sigma}_r^2 = \frac{RMS - MSE}{k}$$

$$\hat{\sigma}_b^2 = \frac{BMS - MSE}{n}.$$

The ANOVA estimator for ρ_2 is

$$\hat{\rho}_2 = \frac{BMS - MSE}{BMS + \dfrac{n}{k}(RMS) + \left(n - 1 - \dfrac{n}{k}\right)MSE}.$$

Under the assumptions of normality, Searle et al. (1992) showed that

$$\frac{(k-1)BMS}{n\sigma_b^2 + \sigma_w^2} \sim X_{(k-1)}^2,$$

$$\frac{(n-1)RMS}{k\sigma_r^2 + \sigma_w^2} \sim X_{(n-1)}^2,$$

and

$$\frac{(n-1)(k-1)MSE}{\sigma_w^2} \sim X_{(n-1)(k-1)}^2.$$

It follows immediately that

$$\text{var}(BMS) = \frac{2(n\sigma_b^2 + \sigma_w^2)^2}{(k-1)},$$

$$\text{var}(RMS) = \frac{2(n\sigma_r^2 + \sigma_w^2)^2}{(n-1)},$$

and

$$\text{var}(MSE) = \frac{2\sigma_w^4}{(n-1)(k-1)}.$$

Saito et al. (2006) used the delta method to derive the asymptotic variance of $\log(\hat{\rho}_2)$ using the approximations

$$\text{var}(\log(X)) = \frac{\text{var}(X)}{E^2(X)},$$

$$\text{cov}(\log(X_1), \log(X_2)) \cong \frac{\text{cov}(X_1, X_2)}{E(X_1) \cdot E(X_2)},$$

$$\text{var}(\log(X_1) - \log(X_2)) \cong \frac{V(X_1)}{E^2(X_1)} + \frac{V(X_2)}{E^2(X_2)} - \frac{\text{cov}(X_1, X_2)}{E(X_1)E(X_2)}.$$

when

$$X_1 = BMS - MSE$$

$$X_2 = BMS + \frac{n}{r}RMS + \left(n - 1 - \frac{n}{k}\right)MSE.$$

Under the assumption that BMS, RMS, and MSE are mutually independent, then

$$\text{var}(X_1) = \text{var}(BMS) + \text{var}(MSE)$$

$$\text{var}(X_2) = \text{var}(BMS) + \frac{n^2}{k^2}\text{var}(RMS) + \left(n - 1 - \frac{n}{k}\right)^2 \text{var}(MSE)$$

and

$$\text{cov}(X_1, X_2) = \text{var}(BMS) - \left(n - 1 - \frac{n}{k}\right)\text{var}(MSE).$$

It can now be easily shown that

$$\text{var}\left(\log(\hat{\rho}_2)\right) = T_1 + T_2 - 2T_3$$

where

$$T_1 = \frac{\text{var}(BMS) + \text{var}(MSE)}{\left\{E(BMS) - E(MSE)\right\}^2}$$

$$T_2 = \frac{\text{var}(BMS) + (n/k)^2\,\text{var}(RMS) + (n - 1 - n/k)^2\,\text{var}(MSE)}{\left\{E(BMS) + (n/k)E(RMS) + (n - 1 - n/k)E(MSE)\right\}}$$

$$T_3 = \frac{\text{var}(BMS) - (n - 1 - n/k)\text{var}(MSE)}{\left[E(BMS) - E(MSE)\right]\left[E(BMS) + (n/k)E(RMS) + (n - 1 - n/k)E(MSE)\right]}.$$

Saito et al. (2006) determined the (n) that minimizes the variance of the estimated ICC, var($\hat{\rho}_2$) when the total number of observations $N = nk$ is fixed. They demonstrated that (n) "which minimizes the variance of ICC varies with $\tau = \sigma_w^2/\sigma_r^2$ and approaches a constant value rapidly with $\tau \to 0$." Note that τ is small when the between raters variance σ_r^2 is relatively much larger than the error variance σ_w^2. The analytic expression for n as derived by Saito et al. (2006) is τ, and at its minimum is

$$n = \sqrt{N} + \sqrt{N}\left[-1 + \frac{2}{\sqrt{N}} - \frac{1}{N} - \frac{2\psi}{\rho_2(N-1)} + \frac{2\psi}{\rho_2(N-1)\sqrt{N}} - \frac{\psi}{\rho_2 N(N-1)}\right] + 0(\tau)$$

$$n = \sqrt{N}\,(1 - \iota)$$

$$= \sqrt{N} + 0(\tau)$$

(2.23)

where $\psi = \sigma_r^2/(\sigma_b^2 + \sigma_r^2 + \sigma_w^2)$, $N = nk$. They concluded that, when N is fixed and the between raters variance σ_r^2 is much larger than the error variance σ_w^2 (i.e., τ is small), $\mathrm{var}(\hat{\rho}_2)$ is minimized if the number of raters and the number of subjects are almost the same.

As an example, when $N = 81$ and $\tau = 0.1$, then $n \cong \sqrt{N}(1 - \tau)$ gives n as 8, while the approximation $n = \sqrt{N}$ gives n as 9.

2.12 Estimation of ICC from Two-Way Random Effects Model with Replicated Measurements

Here, we consider a reliability study in which each of the k subjects is independently measured by each of n raters with equal number of replicates n. We also assume that both subjects and raters are randomly selected from their respective populations. The corresponding model is a balanced two-way random effects model with interaction. The lth measurement on the ith subject by the jth rater is represented by

$$y_{ijl} = \mu + b_i + r_j + \gamma_{ij} + e_{ijl}$$

$$i = 1, 2, \ldots, k; \quad j = 1, 2, \ldots, m, \quad \text{and} \quad l = 1, 2, \ldots, n;$$

where the notations are similar to Case 2. Here, γ_{ij} is the effect of interaction $(rb)_{ij}$ with $\gamma_{ij} \sim N(0, \sigma_\gamma^2)$, and e_{ijl} is the random error with $e_{ijl} \sim N(0, \sigma_e^2)$. We also assume that all the random components of the model are mutually independent. Under these assumptions,

$$\mathrm{var}\left(y_{ijl}\right) = \sigma_b^2 + \sigma_r^2 + \sigma_\gamma^2 + \sigma_e^2,$$

$$\mathrm{cov}\left(y_{ijl}, y_{ij'l'}\right) = \sigma_b^2.$$

Therefore, the ICC measuring the interrater reliability is

$$\rho_4 = \frac{\sigma_b^2}{\sigma_b^2 + \sigma_r^2 + \sigma_\gamma^2 + \sigma_e^2}.$$

(2.24)

It is interpreted as the correlation between randomly selected measurements on the same subject by a pair of randomly selected raters. Moreover,

$$\text{cov}\left(y_{ijl}, y_{ij'l'}\right) = \sigma_b^2 + \sigma_r^2 + \sigma_\gamma^2.$$

Hence, the intraclass correlation that measures intrarater reliability is

$$\rho_5 = \frac{\sigma_b^2 + \sigma_r^2 + \sigma_\gamma^2}{\sigma_b^2 + \sigma_r^2 + \sigma_\gamma^2 + \sigma_e^2}. \tag{2.25}$$

This is the correlation between a pair of randomly selected measurements on the same subject for a randomly selected rater.

We shall estimate both ρ_4 and ρ_5 by equating the mean squares, of the ANOVA table, to their expected values.

Let $SSA = mn\sum_{i=1}^{k}(\bar{y}_{i..} - \bar{y}_{...})^2$, $MSA = SSA/(k-1)$, $SSB = kn\sum_{j=1}^{m}(\bar{y}_{.j.} - \bar{y}_{...})^2$

$$SSE = \sum_{i=1}^{k}\sum_{j=1}^{m}\sum_{l=1}^{n}\left(y_{ijk} - \bar{y}_{ij.}\right)^2$$

$$MSE = SSE/km(n-1)$$

$$MSE = SSB/(m-1)$$

$$SSAB = n\sum_{i=1}^{k}\sum_{j=1}^{m}\left(\bar{y}_{ij.} - \bar{y}_{i..} - \bar{y}_{.j.} + \bar{y}_{...}\right)^2$$

$$MSAB = SSAB/(k-1)(m-1)$$

Here

$$\bar{y}_{...} = \frac{1}{mnk}\sum_{i=1}^{k}\sum_{j=1}^{m}\sum_{l=1}^{n}y_{ijl},$$

$$\bar{y}_{i..} = \frac{1}{mn}\sum_{j=1}^{m}\sum_{l=1}^{n}y_{ijl},$$

$$\bar{y}_{.j.} = \frac{1}{kn}\sum_{i=1}^{k}\sum_{l=1}^{n}y_{ijl},$$

and

$$\bar{y}_{ij.} = \frac{1}{n}\sum_{l=1}^{n} y_{ijl}.$$

From Searle et al. (1992) we have

$$\frac{(k-1)(MSA)}{mn\sigma_b^2 + n\sigma_\gamma^2 + \sigma_e^2} \sim X_{(k-1)}^2,$$

$$\frac{(m-1)(MSB)}{kn\sigma_r^2 + n\sigma_\gamma^2 + \sigma_e^2} \sim X_{(m-1)}^2,$$

$$\frac{(k-1)(m-1)(MSAB)}{n\sigma_\gamma^2 + \sigma_e^2} \sim X_{(k-1)(m-1)}^2,$$

and

$$\frac{km(n-1)MSE}{\sigma_e^2} \sim X_{km(n-1)}^2.$$

Equating the mean squares to their expectations and solving for the variance components, we get

$$\hat{\sigma}_b^2 = \frac{MSA - MSAB}{mn},$$

$$\hat{\sigma}_r^2 = \frac{MSB - MSAB}{nk},$$

$$\hat{\sigma}_\gamma^2 = \frac{MSAB - MSE}{n},$$

and

$$\hat{\sigma}_e^2 = MSE.$$

Moreover, it can be readily shown that

$$\text{var}(MSA) = \frac{2}{(k-1)}\left(mn\sigma_b^2 + n\sigma_\gamma^2 + \sigma_e^2\right)^2,$$

$$\text{var}(MSB) = \frac{2}{(m-1)}\left(kn\sigma_r^2 + n\sigma_\gamma^2 + \sigma_e^2\right)^2,$$

$$\text{var}(MSAB) = \frac{2}{(k-1)(m-1)} \left(n\sigma_\gamma^2 + \sigma_e^2 \right)^2,$$

and

$$\text{var}(MSE) = \frac{2\sigma_e^4}{km(n-1)}.$$

Thus, the variance component estimate of ρ_4 is

$$\hat{\rho}_4 = \frac{k[MSA - MSAB]}{k(MSA) + m(MSAB) + [(k-1)(m-1)](MSAB) + mk(n-1)(MSE)}.$$

Note that, when $n = 1$ (no replications), MSAB and MSE are confounded, and $\hat{\rho}_4$ reduces to $(MSA - MSE/MSA + (m-1)MSE)$, which is identical to the ICC estimator of Case 2. The variance component estimator of ρ_5 may similarly be obtained as

$$\hat{\rho}_5 = \frac{k(MSA) + m(MSB) + (mk - k - m)(MSAB) - mk(MSE)}{k(MSA) + m(MSB) + (mk - k - m)MSAB + mk(n-1)(MSE)}.$$

The large sample variances of $\hat{\rho}_4$ and $\hat{\rho}_5$ may then be obtained only by applying the delta methods, using the exact distributional properties of MSA, MSB, MSAB, and MSE.

2.12.1 Generalizability and Reliability

In basic and applied sciences, experiments are designed to collect measurements, and before applications of the findings one must examine the quality of these measurements. Before applications of the results of such experiments one must keep in mind that the measurements are specific to a particular population. The extent to which a measurement is generalizable to another population depends on the accuracy and the similarity of conditions under which the information has been collected. It was Cronbach et al. (1963) who provided an interpretation of reliability and required that investigators specify the population over which to generalize. The Generalizability Theory (GT) defines measurements as reliable if they allow accurate inference on the population. The question of how to go about designing a study that produces "reliable" measures is the subject of the GT. Cronbach et al. used the terms dependability and generalizability instead of reliability aims at unifying both reliability and validity.

The reason for delaying the discussion on the GT to the end of this chapter is that it is tightly linked to the problem of estimation of components of variations models. To show that the GT is in fact an attempt to decompose the total variations into its respective components, we consider the very simple one-factor experiment.

Let y_{ij} be the observations made on the ith subject under the jth condition. The condition here may be the settings, occasions, or the laboratories involved in the experiment. The generalizability question is: If the observations are made under as similar as possible conditions, can we make conclusions about the population from which they are drawn? Under the one-factor experiment we may assume the two-way random effects model for the observations

$$y_{ij} = \mu + s_i + r_j + e_{ij}.$$

Under the usual assumptions of the random effects model, the total variation in y_{ij} can be decomposed into three sources: $\sigma_{\text{total}}^2 = \text{var}(y_{ij}) = \sigma_s^2 + \sigma_r^2 + \sigma_e^2$, assuming that interaction can be ignored. It has been demonstrated that these components can be estimated from the ANOVA table. The next step is to evaluate the percentage of variation attributed to subjects, raters and error, and these are estimated respectively by, $\hat{\sigma}_s^2/\hat{\sigma}_{\text{total}}^2$, $\hat{\sigma}_r^2/\hat{\sigma}_{\text{total}}^2$, and $\hat{\sigma}_e^2/\hat{\sigma}_{\text{total}}^2$.

Example 2.4

```
data weight;
input packet scale$ weight;
cards;
  1 j 560
  1 p 550
  1 a 575
  2 j 140
  2 p 140
  2 a 140
  3 j 1720
  3 p 1710
  3 a 1715
  4 j 1110
  4 p 1090
  4 a 1100
  5 j 1060
  5 p 1040
  5 a 1035
  6 j 250
  6 p 250
  6 a 251
  7 j 610
  7 p 600
  7 a 610
  8 j 830
  8 p 800
  8 a 815
  9 j 690
```

```
9 p 690
9 a 695
10 j 1210
10 p 1210
10 a 1210
;

/* Two-way random effects model */
proc glm data = weight;
class packet scale;
model weight = packet scale;
random packet scale;
run;
```

Source	Mean square
Packet	665797.867
Scale	258.533
Error	51.867

Source	Type III Expected Mean square
packet	Var(Error) + 3 Var(packet)
scale	Var(Error) + 10 Var(scale)

Hence the variance components estimates are given by (error variance components) = $\hat{\sigma}_e^2 = 51.867$,

(the between-packets variance components)

$$= \hat{\sigma}_b^2 = \frac{665797.867 - 51.867}{3}$$

$$= 221915.3.$$

(the between-scales variance components)

$$= \hat{\sigma}_r^2 = \frac{258.533 - 51.867}{10}$$

$$= 20.67.$$

The total variance is therefore = 221915.3 + 20.67 + 51.867 = 221987.84. The percentage of variations due to the corresponding factors are (51.867/221987.84) × 100 = 0.023% (error percentage), (221915.3/221987.84) × 100 = 99.96% (between packets or the reliability coefficient), and (20.67/221987.84) × 100 = 0.009% (between-scales variation). Clearly, the lowest source of variation is coming from scales.

The above example is quite simple and is presented to demonstrate how elements of the GT can be evaluated. The classical ANOVA and expected MSE components are the essential ingredients of the GT, with the difference that the ANOVA emphasizes the *F*-test, while GT relies on the estimation of variance components.

In summary, assessing raters' reliability is fundamental for the correct interpretation of clinical findings. Generalizability studies, which attempt to determine the sources of variation and their relative contribution to the measurement errors, are very valuable in this respect. In general, improving raters reliability will have a positive impact on the quality of medical and health-related research.

EXERCISES

E2.1. Use the delta method do derive a first-order approximation for the variances of both $\hat{\rho}_4$ and $\hat{\rho}_5$.

E2.2. For the data in Example 2.2, find the numerical values of the ICC $\hat{\rho}_4$ and $\hat{\rho}_5$. The data are given in SAS format in the accompanied CD. The SAS code for this analysis is given by

```
Title 'Computing ICC for replicated trials, with
interaction';
            proc glm;
            class patient method;
        model y= patient| method;
        random patient method
patient*method;
            run;
```

Summary relevant SAS output is given by:

The GLM Procedure

Source	DF	Sum of Squares	Mean Square	F Value	Pr > F
Model	5	2362.033	472.407	631.83	<.0001
Error	54	40.375	0.748		

R-Square	Coeff Var	Root MSE	Y Mean
0.983194	4.956652	0.864688	17.44500

Source	DF	Type III SS	Mean Square	F Value	Pr > F
Patient	2	2288.631	1144.315	1530.48	<.0001
Method	1	55.488	55.488	74.21	<.0001
Patient * Method	2	17.914	8.957	11.98	<.0001

Source	Type III Expected Mean Square
Patient	Var (Error) + 10 Var (Patient * Method) + 20 Var (Patient)
Method	Var (Error) + 10 Var (Patient * Method) + 30 Var (Method)
Patient * Method	Var (Error) + 10 Var (Patient * Method)

E2.3. A large sample $(1 - \alpha)100\%$ confidence on the population ICC, whose estimate may be given by

$$\text{Estimate} \pm Z_{\alpha/2}\sqrt{\text{var}(Estimate)}$$

Construct a 95% confidence on the population parameters ρ_4 and ρ_5.

3

Method-Comparison Studies

3.1 Introduction

In Chapter 2, we discussed models under which reliability studies are conducted. The intraclass correlation coefficient (ICCC) has been the single most important parameter to quantify reliability. The issue of agreement, which is conceptually different from reliability, was discussed briefly in Chapter 1 and the full details of the subject of agreement for categorical data will be discussed in Chapters 5 and 6. Moreover, we have shown that comparing agreements of two faulty methods with one gold standard is equivalent to comparing dependent error variance parameters. The statistical treatment of this problem appeared in the earlier studies by Pitman (1939), Morgan (1939), and Shukla (1973).

In this chapter, our focus is on the issue of agreement between two methods measuring continuous variables. We shall explain the role of regression models with measurement errors in assessing agreement between two measuring instruments. Several data sets will be analyzed using the approach and the results of the seminal papers of Altman and Bland (1983) followed by Bland and Altman (1986, 1999).

3.2 Literature Review

Comparison of two methods of measuring some biological quantity is ubiquitous in biomedical research. This comparison entails the determination of the precision and accuracy of the measuring instruments. The accuracy is defined as the closeness of agreement of the instrument's result, on a given subject, with the true result (gold standard). On the other hand, precision is the closeness of agreement between repeated assessments on the same subject, taken by the same instrument.

Ideally, we should evaluate a method in terms of precision and accuracy. Westgard and Hunt (1973) claimed that precision refers to random errors, whereas accuracy refers to systematic error. Random errors are usually studied first because systematic errors are difficult to evaluate in the presence of random errors. Hence, we study precision and then accuracy. In medicine, method-comparison studies have been analyzed using correlation coefficients. However, correlation measures the strength of association between two methods and is not a measure of agreement for several reasons. First, testing the departure of correlation from zero is meaningless. Since the methods are measuring the same group of subjects, rejecting the null hypothesis of no association does not produce knowledge about the level of agreements between methods. Second, the two methods agree if the scatter plot of measurements taken by one method against the measurements of the second method should be as close as possible to the line of agreement (45° with slope = 1). In other words, the scatter plot should be as close as possible to the line of equality, with slope = 1 and intercept = 0. Another problem with the correlation coefficient, as was noted by Hollis (1996) is its dependence on the range of data. A wider range of data always tends to produce higher correlation. In method-comparison studies if the data include extreme points, for whatever reason, the correlation increases, giving misleading results.

Since we intend to use the regression models in the analysis of agreement, one should note that one of the standard assumptions of simple regression is that the dependent variable Y is measured with error, whereas the independent variable is not. In method-comparison studies we cannot make such an assumption, since both the independent and dependent variables are expected to be measured with error as well. This means that we would have two regressions, depending on the choice of the dependent or the independent variable. One way to overcome this problem is to use techniques of regression analyses that assume errors in both variables, known as Deming (1943) regression. We shall discuss this technique in detail in this chapter. These problems are well documented in the statistical literature and a brief introduction will be given in this chapter. Needless to say that, the papers by Bland and Altman (1986, 1999) and Altman and Bland (1983) outlined alternative approaches that are easily understood by clinicians.

There are several examples in clinical medicines where methods-comparison is needed.

The clinical diagnosis of tricuspid stenosis is often difficult. DE provides an accurate noninvasive tool for the quantitative assessment of valve stenosis. A large volume of data is available in patients with mitral, aortic, and pulmonary stenosis, but a few reports are available on patients with tricuspid stenosis. Fawzy et al. (1989) reported a study that aimed at assessing the accuracy of DE in evaluating tricuspid stenosis compared with cardiac catheterization. Their study was conducted on only 17 patients. They found that Doppler assessment of tricuspid regurgitation compared favorably to that obtained by right ventricular angiography and was not hampered by disturbances of rhythm or possible mechanical interference of valve function that may occur during right ventricular angiography. One criticism that would be directed to this study is the use of regression and correlation to assess the association between two methods, rather than examining the agreement.

Another study was conducted by Liehr et al. (1995) to examine the limits of agreement (LOA) of blood pressures data by comparing oscillometric and intra-arterial blood pressure monitoring methods. They sampled 70 patients who had undergone cardiovascular surgery in the postoperative recovery room. Bland and Altman's approach (BAA) was used in addition to the use of correlation and paired *t*-test. The analyses of the data suggested that the oscillometric measurements method cannot be routinely substituted for the intra-arterial. The measurement methods did not agree.

Altman and Bland (1983) warned that what appears to be strong correlation between two methods of measuring individuals, might have a poor agreement. For example, Serfontein and Jaroszewicz (1978) found a correlation of 0.85 when they compared two methods of assessing gestational ages. They wrongly concluded that because the correlation was high, agreement was good.

3.3 Bland and Altman's Approach

We have already touched on the basic ideas of BAA in Chapter 2. Recall that we assessed the difference between two dependent measures of precision by examining the correlation between the difference and the sum of two sets of measurement errors. This idea is the cornerstone of BAA. According to BAA, the difference of the scores of the two methods is plotted against the average of the scores. In statistical terms, if there is no relationship between the difference and the average, the agreement between the two methods can be summarized using the mean and standard deviation of the differences. The accuracy is then assessed by a test of whether the mean difference is zero. This can be determined using a 95% confidence interval for the difference between the mean scores of the two methods. The LOA are given by

$$\text{Mean difference} \pm 2\,SEM$$

where $SEM = \text{Standard deviation of the difference}/\sqrt{k}$.

Example 3.1

The data are measurements from an agreement study between two raters (radiologist and pathologist) with respect to the angiographic classification of lumen narrowing in the internal carotid artery. There are 107 patients; each patient was assessed twice by the same rater. For the *i*th patient, let (X_{11}, X_{12}) be the two measurements taken by the first rater and (X_{21}, X_{22}) the two measurements taken by the second rater.

```
data reliabil;
input         ID      X11     X12     X21     X22;
cards;
1             23      25      25      26
2             52      71      70      58
3     49      40      59      51
.     .       .       .       .
107   33      23      42      64
      proc means data=reliabil;
      var x11 x12 x21 x22;
      run;
      data new;
      set reliabil;
      *1. Compute the averages of the readings made by
each rater;
            x=(x11+x12)/2;
            y=(x21+x22)/2;
      *2. Compute the sum and the difference of the averages
of the readings;
            u=(x+y);
            v=x-y;

   run;

            * Measure the correlation between u and v;
               proc corr data=new;
                     var u v;
         title;

   run;

      * Retrieve the values of the mean and the standard
deviations of v;

         proc means data=new;
           var v;
           output mean=meanv std=sdv out=reflines;
           run;

data lines;
   set reflines;

      *1. Compute the upper and lower limits of v;
            upper=meanv+ (2*sdv);
            lower=meanv - (2*sdv);
      *2. Convert the variables, upper and lower, into macro
variables;
            call symput('upper', upper);
            call symput('lower', lower);

   run;
```

TABLE 3.1a

Summary Statistics for the Data

Variable	K	Mean	Standard Deviation
X11	107	51.69	26.977
X12	107	52.77	27.44
X21	107	56.79	26.41
X22	107	55.12	25.94

```
* Produce a scatter plot with the upper and lower limits of v;
  goptions reset=all;
  symbol1 v=plus c=blue;
  proc gplot data=new;
  plot v*u / vref=&upper &lower lvref=2
                cvref=red;
run; quit;

/* To evaluate the concordance correlation between the two
raters, we need to evaluate the means, variances of x and y,
and the Pearson's correlation. */
data concord;
set new;
  proc means data=concord;
     var x y; run;
proc corr data=concord;
var x y;   run;
```

SAS OUTPUT (see Tables 3.1a and b)
Therefore, the estimated CCC between x and y is

$$\hat{\rho}_c = \frac{2(0.939)(26.5)(26.8)}{(26.5)^2 + (26.8)^2 + (3.73)^2} = 0.93.$$

BAA graphical limits of agreement are given in Figure 3.1.

TABLE 3.1b

Simple Statistics for the Variables u and v

Variable	K	Mean	Standard Deviation
u	107	108.19	51.51
v	107	−3.73	9.14

Pearson correlation coefficients between u and v = 0.08, p-value = 0.407

x	107	52.23	26.52
y	107	55.96	25.79

Pearson correlation coefficients between x and y = 0.939.

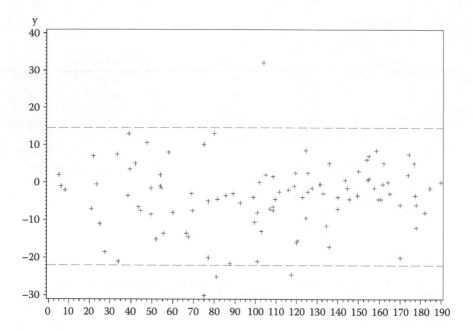

FIGURE 3.1
Bland and Altman's limits of agreement for Example 3.1 data.

Remark

If there appears to be a relationship between the difference and the mean, then the LOA will not be acceptable, since the range of the difference is related to the average value. This has been formalized by a procedure due to Bradley and Blackwood (1989).

3.4 Bradley–Blackwood Procedure

Let $U_i = Y_{i1} - Y_{i2}$, and $V_i = (Y_{i1} + Y_{i2}/2)$, where Y_{i1}, Y_{i2} are the readings made by the two methods on the ith subject. The Bradley–Blackwood Procedure (BBP) fits a regression of U_i and V_i. That is,

$$U_i = \alpha + \beta V_i$$

where α and β, the intercept and slope, respectively, are given by

$$\alpha = \mu_1 - \mu_2 - \frac{1}{2}\left[\frac{\sigma_1^2 - \sigma_2^2}{\sigma_v^2}\right]\left[\frac{\mu_1 + \mu_2}{2}\right] \tag{3.1}$$

and

$$\beta = \frac{1}{2}\left[\frac{\sigma_1^2 - \sigma_2^2}{\sigma_v^2}\right] \qquad (3.2)$$

where $\mu_j = E(Y_{ij})$, $\sigma_j^2 = \text{var}(Y_{ij})$, and $\sigma_v^2 = \text{var}(V_i)$.

Since $\mu_1 = \mu_2$ and $\sigma_1^2 = \sigma_2^2$, if and only if $\alpha = \beta = 0$, a simultaneous test of the equivalence of the means and variances of Y_1 and Y_2 is an F test, calculated from the results of a regression of D and U. The test statistic is

$$F_{2,k-2} = \left(\sum_{i=1}^{k} U_i^2 - SS_{reg}\right)\Big/\left(2MS_{reg}\right) \qquad (3.3)$$

where SS_{reg} and MS_{reg} are, respectively, the residual sum of squares and the mean-square with $k - 2$ degrees of freedom, from the regression of D on U. The above quantities may be readily obtained from the output of the ANOVA regression available in SAS. The ANOVA Table 3.2 takes the form.

Note that

$$S_v^2 = \sum_{i=1}^{k} V_i^2 - \left(\sum_{i=1}^{k} V_i\right)^2\Big/k \qquad (3.4)$$

$$S_u^2 = \sum_{i=1}^{k} U_i^2 - \left(\sum_{i=1}^{k} U_i\right)^2\Big/k. \qquad (3.5)$$

Equations 3.4 and 3.5 are the corrected sum of squares around the respective means of V_i and U_i.

Bartko (1994) showed that the ICC may be directly obtained from the ANOVA regression Table 3.2 as

$$\hat{\rho}_1 = \frac{4S_v^2 - S_u^2}{4S_v^2 + S_u^2}. \qquad (3.6)$$

Moreover, testing the quality of correlated variances (precisions) is equivalent to testing $H_O : \beta = 0$.

Example 3.2

The following data are taken from the discipline of cardiology. Two methods are used to measure the AVA and we are interested in assessing the agreement between the two methods.

TABLE 3.2

Regression ANOVA for Methods-Comparison

Source	Df	SS	Regression Quantity
Between-subjects	$k-1$	BSS	$2S_v^2$
Within-subject	k	WSS	$1/2\sum U_i^2$
Between methods (raters)	1	RSS	$(\Sigma U_i)^2/2k$
Residual	$k-1$	SSE	$1/2S_U^2$
Total	$2k-1$		

```
data methods;
input patient method1 method2;
cards;
1     0.8     0.8
2     0.7     0.7
3     1       1
4     0.8     0.9
5     0.9     0.9
6     0.7     0.8
7     1       1.1
8     0.8     0.7
9     0.9     0.9
10    0.6     0.7
11    0.7     0.5
12    1       1
13    0.7     0.7
14    0.6     0.7
15    0.7     0.7
16    0.6     0.7
17    0.4     0.4
18    0.5     0.6
19    0.4     0.4
20    0.6     0.6
21    0.3     0.3
22    0.5     0.5
23    0.3     0.3
24    0.7     0.7
25    0.6     0.5
26    0.6     0.6
27    0.8     0.9
28    0.7     0.8
29    0.6     0.6
30    0.5     0.4
31    0.3     0.3
32    0.8     0.8
33    0.8     0.7
```

```
34   0.5    0.4
35   0.5    0.4
36   0.8    0.7
37   0.6    0.6
38   1      1.1
39   0.9    0.8
40   1      1
;
proc means data=methods;
var method1 method2;
run;


proc gplot;
     /* Define symbol characterictics */
symbol1 interpol=r
        value=plus
            height=3
        cv=rd
        ci=blue
        width=2;
/* generate scatter plot with regression */
plot method1*method2 /regeqn;
run;
proc corr data=methods;
var method1 method2;
run;
data new;
set methods;
u=method1-method2;
v=(method1+method2)/2;
proc reg data=new;
model u=v;
run;
proc univariate data=new;
var u v;
run;

/* To verify that the Bradly-Blackwood approach
gives ICC as indicated by Bartko*/
data new;
set new;
rater=1; score=method1; output;
rater=2; score=method2; output;
proc glm data=new;
class patient;
model score= patient;
random patient;
run;
```

Since the intercept is far from zero, as can be seen from Figure 3.2, one may conclude that the two methods are biased relative to each other, even though the slope is not significantly different from unity.

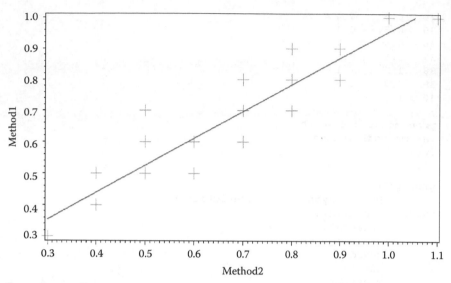

Regression equation:
method = 0.097452 + 0.856688*method2

FIGURE 3.2
Scatter plot showing the line of regression.

Variable	Mean	Standard Deviation
Method1	0.68	0.200
Method2	0.68	0.220

Pearson correlation coefficient is $r = 0.94$. Therefore, the CCC is

$$\hat{\rho}_c = \frac{2(0.94)(0.2)(0.22)}{0.04 + 0.0484} = 0.936.$$

The closeness of the CCC to Pearson's correlation coefficient is due to the fact that the two methods have the same mean and almost the same standard deviations (see Table 3.3a).

TABLE 3.3a
ANOVA to Test the Significance of the Regression Equation

Source	DF	Sum of Squares	Mean Square	F Value	Pr > F
Model	1	0.015	0.015	2.85	0.0997
Error	38	0.205	0.005		

TABLE 3.3b

SAS ANOVA to Obtain the ICC

Source	DF	Sum of Squares	Mean Square	F Value	Pr > F
Model	39	3.338	0.0856	31.12	<0.0001
Error	40	0.11	0.00275		
Source	DF	Type III SS	Mean Square	F Value	Pr > F
Patient	39	3.33800000	0.08558974	31.12	<0.0001
Source	Type III Expected Mean square				
Patient	Var (Error) + 2 Var (patient)				

The nonsignificance of the regression equation p-value = 0.0997 means that the hypothesis of equality of means and variances of the two methods is justified by the data. Moreover, the correlation between U and V is low (approximately 0.26). This is an indication that the LOA between the two methods are acceptable.

The ANOVA estimator of the ICC can be obtained as in Chapter 2: $\hat{\sigma}_e^2 = 0.00257$, while the between-patients variance component estimator is $\hat{\sigma}_b^2 = (0.0856 - 0.00275)/2 = 0.0414$. Therefore, the estimated reliability coefficient is $\hat{\rho} = 0.0414/(0.0414 + 0.00275) = 0.937$. The results are obtained from the GLM output of the attached SAS program (see Table 3.3b).

To obtain the ICC using Bartko's expression, we found $S_u^2 = 0.22$, $S_v^2 = 1.669$. Hence, from Equation 3.6,

$$\hat{\rho}_1 = \frac{4(1.669) - 0.22}{4(1.669) + 0.22} = 0.936.$$

There is no doubt that statistical testing is useful in the analysis of method-comparison studies. These tests help in determining the precision and accuracy in order to objectively judge the acceptability of different methods as tools of medical decision-making. Regression methods cannot provide us with definite answers regarding the issue of acceptability. However, they provide estimates of the magnitude of different types of errors. As we have already indicated, Pearson's correlation provides no information about the magnitude of error. Provided that the relationship between the measurements made by the two methods is linear, the least-square method of estimation produces useful results. One may achieve linearity by employing a monotonic transformation, but the results should be cautiously interpreted. Finally, while the ANOVA of the linear regression of the difference on the mean provides a simultaneous test of equality of the two means and the two variances, it does not provide specific estimates of the measurement errors.

In the next section, we shall discuss the modifications introduced to the method of least-squares to estimate regression parameters under different structures of measurement errors.

3.5 Regression Methods for Methods Comparison

We start this section with a short introduction on the measurement error model.

Recall the simple linear regression model, written as

$$y_i = \alpha + \beta x_i + e_i, \quad i = 1, 2, \ldots, n.$$

A fundamental assumption in the analysis of the simple linear regression model is that $(x_1, x_2, \ldots, x_n)$ are fixed, or in other words, measured without error. It is well known that the least-squares estimator of β is given by

$$\hat{\beta} = \frac{S_{xy}}{S_x^2},$$

where

$$S_x^2 = \sum_{i=1}^{n}(x_i - \bar{x})^2, \quad S_{xy} = \sum_{i=1}^{n}(x_i - \bar{x})(y_i - \bar{y}).$$

The estimator $\hat{\beta}$ is unbiased to β, and $\hat{\alpha} = \bar{y} - \hat{\beta}\bar{x}$ is unbiased for α.

Now suppose that $(x_1, x_2, \ldots, x_n)$ are random examples from a normal distribution so that

$$E(x_i) = \mu_x, \quad \text{var}(x_i) = \sigma_x^2.$$

It is interesting to note that even for the case when $(x_1, x_2, \ldots, x_n)$ are not fixed, $(\hat{\beta}, \hat{\alpha})$ are still unbiased estimators for (β, α).

Now suppose that instead of observing x_i, one observes the sum

$$X_i = x_i + u_i$$

where

$$u_i \sim N(0, \sigma_u^2).$$

Fuller (1987) differentiated between two models, the first is when x_i is fixed and the resulting model is termed "Functional Model," and the second is when x_i is random and the resulting model is named "Structural Model."

Let us consider the case where x_i is random. We shall assume that (x_i, e_i, u_i) are independently and normally distributed random variables with mean vector (μ_x, 0, 0) and for the time being, a diagonal variance matrix where the diagonal elements are (σ_x^2, σ_e^2, σ_u^2); consequently, we have

$$E(X_i) = \mu_x,$$

$$\text{var}(X_i) = \sigma_x^2 + \sigma_u^2,$$

$$E(y_i) = \alpha + \beta\mu_x,$$

$$\text{var}(y_i) = \beta^2\sigma_x^2 + \sigma_e^2,$$

$$\text{cov}(X_i, y_i) = E[\{\beta(x_i - \mu_x) + e_i\}\{x_i - \mu_x + \mu_i\}] = \beta\sigma_x^2.$$

Denote the naive least-squares estimator of β by

$$\hat{\gamma} = \left\{\sum_{i=1}^{n}(X_i - \overline{X})^2\right\}^{-1}\sum_{i=1}^{n}(X_i - \overline{X})(y_i - \overline{y}).$$

Taking the expectations of both sides and using the results imposed on the distribution of the random vector (x_i, e_i, u_i), we get

$$E(\hat{\gamma}) = \frac{1}{\sigma_x^2 + \sigma_u^2} \cdot \beta\sigma_x^2$$

$$= \frac{\sigma_x^2}{\sigma_x^2 + \sigma_u^2} \cdot \beta = (\rho_{xu})\beta.$$

This expression is correct as long as (x_t, e_t, u_t) are independently distributed. The quantity $\rho_{xu} < 1$ is the familiar expression for the ICC and in fact is called the "reliability ratio" (see Fuller, 1987). The conclusion is that the error in X_i caused the least-squares estimator to be biased toward zero, or *attenuated*.

There are many situations where measurement error is likely to occur. For example, it is well known that systolic blood pressure (SBP) can be used as a predictor for coronary heart disease (CHD). It is also well known that SBP is prone to measurement error. To increase the accuracy of the measured SBP levels, we may either replicate the measurements or use two machines. We then have a situation when both sets of measurements of SBP are with errors. Another example from case-control studies of disease and serum hormone levels, where measurement error would be of major concern. The sources of error would be either from the within-subject variation or the laboratory errors. Another example is from environmental epidemiology where house-hold exposure to lead levels is of primary concern. Here the accuracy of the measuring devices and the location and the time at which lead levels measure-ments are taken are major concern. As a matter of fact this is an area applica-tion where errors can be correlated, making the data analysis more complex.

There are two important situations where the classical measurements error model has very useful applications. The first case is when the ratio $\lambda = \sigma_e^2/\sigma_u^2$ is known and the errors e and u are uncorrelated, and the second case is when the ratio $\lambda = \sigma_e^2/\sigma_u^2$ is known but the errors are correlated, that is, when corr$(u,e) \neq 0$.

In the previous discussion by Bartko (1994), the comparison between two methods clearly involves the estimation of a linear relationship between the two sets of measurements. Linnett (1990) and Barnett (1969) indicated that if the line of relationship between x and y does not pass through the origin or if the slope of the line deviates significantly from unit, a proportional sys-tematic error exists between the measurements of the two methods. In this case, the method of ordinary least-squares would be inappropriate. This is because this method presupposes that one of the methods is without ran-dom measurement error and that the standard deviation of the error distri-bution of the other method is constant throughout the measurement range (see Snedecor and Cochran, 1980). These assumptions are never satisfied in many practical situations, because both methods are usually subject to ran-dom measurement errors, and the standard deviations or the variances of the error distributions are in many biological data, more likely proportional to the variable level than constant (Linnett, 1990). Ignoring these facts are the main reason for what is known as attenuation or downwards bias of the slope estimate. The problem is dealt with by using a regression model that takes into account random measurement errors for both methods. We shall distinguish between two situations. The first is when the relationship between measurements of two methods is linear with constant errors, and the second is when the measurement errors, of the two methods are proportional.

3.5.1 Constant Errors

Let us assume that we observe

$$x_i = X_i + \eta_i \quad \text{and} \quad y_i = Y_i + \xi_i.$$

We assume that X_i and Y_i are the true values; η_i and ξ_i are the correspond-ing measurement errors. We also assume that the unobservable quantities X_i and Y_i are linearly related such that $Y_i = \alpha + \beta X_i$, and assume that $E(\eta_i) = E(\xi_i) = 0$, the errors (η_i, ξ_i) are uncorrelated with each other and are uncorrelated with the true unobserved values (X_i, Y_i). Moreover, we assume that the observations are independent, $\text{var}(X_i) = \sigma_x^2, \text{var}(\eta_i) = \sigma_\eta^2,$ $\text{var}(\xi_i) = \sigma_\xi^2.$

Since $y_i = \alpha + \beta(x_i - \eta_i) + \xi_i$

$$\text{var}(y_i - \alpha) = \beta^2 \text{var}(x_i - \eta_i) + \text{var}(\xi_i)$$

$$= \beta^2\sigma_x^2 + \sigma_\xi^2.$$

Assuming that $\lambda = \sigma_\xi^2/\sigma_\eta^2$ is known (Deming's regression). For a given λ, and assuming the normality of X_i, η_i, ξ_i, the MLE was given by Linnet (1990) as

$$\hat{\beta} = \frac{S_y^2 - \lambda S_x^2 + \left\{\left[S_y^2 - \lambda S_x^2\right]^2 + 4\lambda S_{xy}^2\right\}^{1/2}}{2S_{xy}} \tag{3.7}$$

$$\hat{\alpha} = \bar{y} - \hat{\beta}\bar{x}$$

where

$$S_y^2 = \sum_{i=1}^{k}(y_i - \bar{y})^2, \quad S_x^2 = \sum_{i=1}^{k}(x_i - \bar{x})^2, \quad \text{and} \quad S_{xy} = \sum_{i=1}^{k}(x_i - \bar{x})(y_i - \bar{y}).$$

The estimates of the regression parameters were first derived by Kummel (1879) but are usually attributed to Deming (1943).

Note that the crucial point in the above analysis is whether we can be certain about the value of λ. Carroll and Ruppert (1996) questioned whether the assumptions about λ are valid. They concluded that an invalid λ leads to biased estimates of α and β. It is suggested that an estimator of λ may be obtained either from the data such that

$$\hat{\lambda} = \left(\sum_{i=1}^{k}\left(x_i - \frac{x_i + y_i}{2}\right)^2 + \left(y_i - \frac{x_i + y_i}{2}\right)^2\right)\Big/k = \sum_{i=1}^{k}\frac{(x_i - y_i)^2}{2k},$$

or from the error variance estimator provided by the method, which can be done if each method provides replications of each patient or subject. Approximation of the estimated standard errors of the regression parameters estimates were given by Strike (1995) as

$$SE(\hat{\beta}) = \left\{\hat{\beta}^2\left(\frac{1 - r^2}{r^2(k - 2)}\right)\right\}^{1/2},$$

and

$$SE(\hat{\alpha}) = \left\{ (SE(\hat{\beta}))^2 \frac{(S_{xx} + k\bar{x}^2)}{k} \right\}^{1/2}.$$

Example 3.3

The following data represent hypothetical measurements taken by two methods on a sample of 11 subjects:

1	31	20
2	40	28
3	17	12
4	101	98
5	106	104
6	47	47
7	11	13
8	104	93
9	57	57
10	100	87
11	40	43

It was found that $\hat{\lambda} = 27.55$, $\hat{\beta} = 0.954(0.0524)$, $\hat{\alpha} = -1.925(3.6)$, and $r = 0.987$. It should be noted that these regression estimates and their standard errors are quite close to the regression estimates when measurement errors are ignored.

3.5.2 Proportional Errors

Linnet (1990) examined the situation when the error variances are assumed proportional to the square of the average of the true values. That is,

$$\sigma_\eta^2 = V(\eta_i) = h_x^2 \left[\frac{X_i + Y_i}{2} \right]^2 \qquad (3.8)$$

$$\sigma_\xi^2 = V(\xi_i) = h_y^2 \left[\frac{X_i + Y_i}{2} \right]^2. \qquad (3.9)$$

Under this assumption $\lambda = (h_x^2/h_y^2)$ is constant, he suggested that a modified estimate of the slope may be obtained upon minimizing

$$Q_w = \sum_{i=1}^{k} \left[w_i \left(x_{i'} - \hat{X}_i \right)^2 + \lambda w_i \left(y_i - \hat{Y}_i \right)^2 \right] \qquad (3.10)$$

where $w_i = \left[(x_i + y_i)/2 \right]^{-2}$ and $\hat{Y}_i = \alpha_w + \beta_w \hat{X}_i$.

Denoting

$$\bar{x}_w = \frac{\sum w_i x_i}{\sum w_i},$$

$$\bar{y}_w = \frac{\sum w_i y_i}{\sum w_i},$$

$$S_{xw}^2 = \sum w_i (x_i - \bar{x}_w)^2,$$

$$S_{yw}^2 = \sum w_i (y_i - \bar{y}_w)^2,$$

$$S_{xyw} = \sum w_i (x_i - \bar{x}_w)(y_i - \bar{y}_w).$$

Linnet showed that the slope estimate is

$$\hat{\beta}_w = \frac{\left[S_{yw}^2 - \lambda S_{yw}^2 + \left\{ \left(S_{yw}^2 - \lambda S_{yw}^2 \right)^2 + 4\lambda S_{xyw}^2 \right\}^{1/2} \right]}{2\lambda S_{xyw}} \tag{3.11}$$

and

$$\hat{\alpha}_w = \bar{y}_w - \hat{\beta}_w \hat{x}_w. \tag{3.12}$$

Note that the weights are iterative and the starting weights are

$$w_i = \left[\frac{x_i + y_i}{2} \right]^{-2} \tag{3.13}$$

where x_i and y_i are the observed measurements. The proportionality factor λ is estimated by

$$\hat{\lambda} = \sqrt{\frac{\sum_{i=1} w_i (y_i - \hat{y})^2}{k - 2}}$$

The standard errors of the estimates are given by

$$SE(\hat{\beta}_w) = \left(\frac{\hat{\lambda}}{S_{xw}} \right)$$

and

$$SE(\hat{\alpha}_w) = \hat{\lambda} \left(\frac{1}{\sum\limits_{i=1} w_i} + \frac{(\bar{x}_w)^2}{S_{xw}} \right)^{1/2}.$$

3.6 Correlated Measurement Errors

Models with measurement errors were discussed in the previous sections under the assumptions that the errors were stochastically independent. There are situations however, when these errors are correlated, adding to the complexity of what already seems to be a complicated situation that renders the ordinary least-squares method invalid. Carroll et al. (1995) examined the consequences of ignoring measurement error. Schaalje and Butts (1993) provided a lucid discussion to explore the effect of correlated measurement errors on the method of least-squares. They started with the model

$$Y_i = a + bX_i + \epsilon_i \tag{3.14}$$

where Y_i and X_i are the unobservable true measurements made by the two methods, $\epsilon_i \sim N(0, \sigma^2)$ and $X_i \sim N(\mu_x, \sigma_x^2)$. It follows directly that

$$E(Y_i) = a + b\mu_x,$$

and

$$\text{var}(Y_i) = b^2 \sigma_x^2 + \sigma^2.$$

Now, as before, we assume that (x_i, y_i) are observed and measured with errors so that

$$x_i = X_i + \eta_i$$

$$y_i = Y_i + \xi_i \tag{3.15}$$

with $\eta_i \sim N(0, \sigma_\eta^2)$, $\xi_i \sim N(0, \sigma_\xi^2)$.

The added assumption to this structure is that

$$\text{Corr}(\eta_i, \xi_i) = c \neq 0.$$

One can easily show

$$E(x_i) = \mu_x, \quad E(y_i) = a + b\mu$$

$$\text{var}(x_i) = \sigma_x^2 + \sigma_\eta^2, \quad \text{var}(y_i) = b^2\sigma_x^2 + \sigma^2 + \sigma_\xi^2.$$

Multiplying both sides of Equation 3.14 by X_i and taking expectation with respect to the distributions of the random variables, we get

$$E(X_iY_i) = a\mu_x + b(\mu_x^2 + \sigma_x^2).$$

Hence,

$$\text{cov}(X_i, Y_i) = b\sigma_x^2,$$

and

$$\gamma^2 = \overset{2}{\text{Corr}}(X_i, Y_i) = \frac{b^2\sigma_x^2}{(b^2\sigma_x^2 + \sigma^2)}.$$

One can also show from Equation 3.15 that

$$\text{cov}(x_i, y_i) = b\sigma_x^2 + E(\xi_i\eta_i)$$
$$= b\sigma_x^2 + c\sigma_\eta\sigma_\xi.$$

The joint distribution of (x_i, y_i) is clearly a bivariate normal with mean vector

$$E\begin{pmatrix} x_i \\ y_i \end{pmatrix} = (\mu_x \quad a + b\mu_x)',$$

and variance–covariance matrix

$$\Sigma = \begin{pmatrix} \Sigma_{11} & \Sigma_{12} \\ \Sigma_{21} & \Sigma_{22} \end{pmatrix}$$

where

$$\Sigma_{11} = \text{var}(x_i) = \sigma_x^2 + \sigma_\eta^2$$

$$\Sigma_{12} = \Sigma_{21} = b\sigma_x^2 + c\sigma_\eta\sigma_\xi$$

$$\Sigma_{22} = \text{var}(y_i) = b^2\sigma_x^2 + \sigma^2 + \sigma_\xi^2.$$

From the properties of the bivariate normal distribution (see Chapter 1), the conditional distribution of y_i, given x_i is a univariate normal, with conditional expectation

$$E(y_i|x_i) = E(y_i) + \overset{-1}{\Sigma_{11}}\Sigma_{12}(x_i - \mu_x)$$

$$= a + b\mu_x + \frac{b\sigma_x^2 + c\sigma_\eta\sigma_\xi}{\sigma_x^2 + \sigma_\eta^2}(x_i - \mu_x)$$

and conditional variance

$$\text{var}(y_i|x_i) = \Sigma_{22} - \Sigma_{22}\overset{-1}{\Sigma_{11}}\Sigma_{12}$$

$$= b^2\sigma_x^2 + \sigma^2 + \sigma_\xi^2 - \frac{\left(b\sigma_x^2 + c\sigma_\xi\sigma_\eta\right)^2}{\sigma_x^2 + \sigma_\eta^2}.$$

Equivalently, we write

$$y_i|x_i \sim N\left(a + b\mu_x + b_o(x_i - \mu_x), \sigma_o^2\right) \tag{3.16}$$

where

$$b_o = \frac{b\sigma_x^2 + c\sigma_\xi\sigma_\eta}{\sigma_x^2 + \sigma_\eta^2}. \tag{3.17}$$

$$\sigma_o^2 = \sigma^2 + b^2\sigma_x^2 + \sigma_\xi^2 - \frac{\left(b\sigma_x^2 + c\sigma_\eta\sigma_\xi\right)^2}{\sigma_x^2 + \sigma_\eta^2}, \tag{3.18}$$

and

$$\gamma_o^2 = \overset{2}{\text{Corr}}(x_i, y_i) = \frac{\left(b\sigma_x^2 + c\sigma_\eta\sigma_\xi\right)^2}{\left(\sigma_x^2 + \sigma_\eta^2\right)\left(b^2\sigma_x^2 + \sigma_\xi^2 + \sigma^2\right)}. \tag{3.19}$$

Suppose, now that error measurement is present but ignored. In this case, one would fit the straight line

$$y_i = a_o + b_o x_i + e_{oi} \tag{3.20}$$

where $e_{oi} \sim N(0, \sigma_o^2)$.

The ordinary least-squares estimate $\hat{b}_o$ is an unbiased estimator of b_o, the residual mean square $\hat{\alpha}_o^2$ is an unbiased estimator of α_o^2, and the coefficient of determination $\hat{\gamma}_o^2$ is an unbiased estimator of γ_o^2 (Schaalje and Butts, 1993).

Following them, we write

$$\beta = c\frac{\sigma_\xi}{\sigma_\eta}, \quad \Delta = b - \beta.$$

Substituting into Equations 3.17 and 3.18, we get

$$E\left(\hat{b}_o\right) = b - \frac{\Delta\sigma_\eta^2}{\sigma_x^2 + \sigma_\eta^2}, \tag{3.21}$$

$$E\left(\hat{\sigma}_o^2\right) = \sigma^2 + \left(1 - c^2\right)\sigma_\xi^2 + \frac{\Delta^2\sigma_x^2\sigma_\eta^2}{\sigma_x^2 + \sigma_\eta^2}, \tag{3.22}$$

$$E\left(\hat{\gamma}_o^2\right) = \gamma^2\left(1 + \frac{\beta\sigma_\eta^2}{b\sigma_x^2}\right)^2\left(1 + \frac{\sigma_\eta^2}{\sigma_x^2}\right)^{-1}\left(1 + \frac{\sigma_\xi^2}{\sigma_y^2}\right)^{-1} \tag{3.23}$$

where

$$\sigma_y^2 = \sigma^2 + b^2\sigma_x^2.$$

Schaalje and Butts (1993) established the following:

1. From Equation 3.21, if we set $c = 0$, then $\Delta = b$ and

$$E(\hat{b}_o) = b\left[\frac{\sigma_x^2}{\sigma_x^2 + \sigma_\eta^2}\right]. \tag{3.24}$$

The ratio $R = (\sigma_x^2/\sigma_x^2 + \sigma_\eta^2) < 1$ is known as the measurement reliability. The fact that $E(\hat{b}_o) < b$ is known as the attenuation is due to error measurement.

2. Again, if $c = 0$

$$E(\hat{\sigma}_o^2) = \sigma^2 + \sigma_\xi^2 + b^2\left(\sigma_x^{-2} + \sigma_\eta^{-2}\right). \tag{3.25}$$

That is, $E(\hat{\sigma}_o^2)$ exceeds $\sigma^2 + \sigma_\xi^2$ by $b^2\left(\sigma_x^{-2} + \sigma_\eta^{-2}\right)$.

3. For $c \neq 0$, based on Equation 3.23, the coefficient of determination (γ^2) increases due to the presence of measurement error. For example, if one takes the extreme case; $c = 1$ and $\beta = b$, we get

$$E(\hat{\gamma}_o^2) = \gamma^2\left[1 + \frac{\sigma^2\sigma_\eta^2}{\sigma_x^2\left(\sigma^2 + b^2\sigma_x^2 + b^2\sigma_\eta^2\right)}\right] > \gamma^2. \tag{3.26}$$

Most of the method-comparison studies that we have discussed in the above sections depend on regression analysis. What should be known is that there are certain assumptions that should be satisfied for the regression analysis to yield reliable results. The first is that the relationship between the readings of the two methods must be linear. The scatter plot may reveal whether there is a linear relationship or not. Any curvature, the data may invalidate the conclusions based on the assumption of linearity.

The variable that has been identified as the X or the independent variable is assumed to be measured without error. In method-comparison studies, both methods are fallible, that is measurements by both methods are subject to errors. Therefore, one has to accept the fact that the variable designated as "X" will be measured with error. Another assumption is that the measurements labeled as "Y," should be normally distributed so that the statistic used to test the departure of the intercept from zero and the deviation of the slope from one, has the right distribution (either the *t*- or the *F*-distribution).

Moreover, the assumption of constancy of the variance of the Y variable for every value of X must be satisfied as well. This is known as the homoscedasticity assumption. Scatter plots may show the violation of this assumption by the data. For example, if the variation of Y increases as the values of X increase, this should be taken as evidence against variance homogeneity.

As a final remark, attention must be given to outlying observations. Linnett (1999) indicated that a single point that is away from the scatter plot exerts a great influence on the estimated intercept and slope. Removing what is believed to be an outlying observation might be seen as "data tempering." Replicating measurements may minimize the problem.

The above cautionary remarks are meant to help in the analysis of methods-comparison using regression. An improved analysis depends on other factors such as the skill, the knowledge, and the experience of the analyst.

3.7 Assessing Agreement in Methods-Comparison Studies with Replicate Measurements

3.7.1 Replicates Are Exchangeable

It should be emphasized that the fundamental idea of the methods-comparison is to examine the average difference between the methods and among the subject variations. When we look at the data of Example 3.1, we find that we have two replicates for each rater (method). Bland and Altman (1999) presented several techniques to deal with this situation.

In many applications it might be useful to obtain replicate measurements by each method on each individual so that the repeatability of the two methods can be compared. The basic idea of the LOA methods as described by Bland and Altman (1986) is to examine the difference between methods, and also consider the variability in those methods across the subjects in the study. The authors implicitly assumed that the difference between the two methods is constant across the range of measurements. There are two different situations to consider for replicated observations. Bland and Altman (2007) described two situations, the first being when observations for the same subject are considered as a series of measurements for a biological quantity that does not vary during the observation period. An example is measurements of carotid artery stenosis taken on the same day. The second situation is when the quantity of interest is unstable, for example, blood pressure. The distinction between the two situations has direct implications on the study design. In the first situation, there is no need to acquire the same number of replicates, while in the second situation the same number of replicates would be required.

These techniques depend on hand calculations. Recently, Carstensen et al. (2008) used the linear mixed to estimate the levels of agreement among methods with repeated (replicated) measurements. The linear mixed model representation of this situation is

$$y_{ijl} = \mu_j + s_i + c_{ij} + \varepsilon_{ijl}.$$

Here y_{ijl} is the lth reading $(l = 1, 2, \ldots, m)$ made on the ith subject $(i = 1, 2, \ldots, k)$ by the jth method, $(j = 1, 2, \ldots, n)$. Furthermore, it is assumed that

$$c_{ij} \sim N\left(0, \tau_n^2\right), \quad \varepsilon_{ijl} \sim N\left(0, \sigma_n^2\right).$$

The variance component τ_n^2 of the interaction term C_{ij} measures the between-subjects variation for the nth rater, while σ_n^2 measures the within-subject variation. For estimability, we assume that $\tau_1^2 = \tau_2^2 = \cdots = \tau_n^2$.

Suppose that we have two methods. Therefore, the variance of the difference between pairs of measurements is

$$\operatorname{var}\left(y_{i1} - y_{i2}\right) = 2\tau^2 + \sigma_1^2 + \sigma_2^2. \tag{3.27}$$

Therefore, the LOA are estimated by

$$\hat{\mu}_1 - \hat{\mu}_2 \pm 2\sqrt{2\hat{\tau}^2 + \hat{\sigma}_1^2 + \hat{\sigma}_2^2}.$$

Carstensen et al. (2008) provided an SAS code to estimate the relevant quantities. The first requirement to implement the code is to set the data in a univariate structure, as in Table 3.4.

TABLE 3.4

Pasteurella haemolytica Bacterial Count Readings Made by Two Raters and Three Replicates

1	1	1	52
1	2	1	62
1	3	1	60
1	1	2	72
1	2	2	93
1	3	2	78
2	1	1	68
2	2	1	62
2	3	1	64
2	1	2	68
2	2	2	63
2	3	2	65
.	.	.	.

Note: Data are given for the first two slides. The rest of the data are in the accompanied CD.

3.7.2 Replicates Are Not Exchangeable

Suppose that replicates are taken across different settings; for example, in time sequence. This situation precludes the exchangeability of the replicates and this requires modification of the random effects model, so that

$$y_{ijl} = \mu_j + s_i + \gamma_{il} + C_{ij} + e_{ijl},$$

$$\gamma_{il} \sim N\left(0, \sigma_w^2\right), \quad c_{ij} \sim N\left(0, \tau_n^2\right), \quad e_{ijl} \sim N\left(0, \sigma_n^2\right).$$

Note that $n = 1, 2$, $\tau_1^2 = \tau_2^2$.

$$\text{var}\left(y_{i1} - y_{i2}\right) = 2\tau^2 + \sigma_1^2 + \sigma_2^2$$

and the LOA are

$$\hat{\mu}_1 - \hat{\mu}_2 \pm 2\sqrt{2\hat{\tau}^2 + \hat{\sigma}_1^2 + \hat{\sigma}_2^2}.$$

Example 3.4

Bacterial counts in swabs from the respiratory tract of 15 cows are repeated by two trained veterinarians (raters). Each rater provided three readings from the same slide. We are using these data to determine the level of agreement between the two raters under the two discussed situations.

Carstensen et al. (2008) provided the following SAS codes:

```
/* Replicates are exchangeable*/

proc mixed;
class slide rater;
model count = rater slide /s;
random rater*slide;
repeated slide /group=rater;
run;

/* Replicates are NOT exchangeable*/

proc mixed;
class slide rater repl;
model count = rater slide /s;
random rater*slide slide*repl;
repeated slide /group=rater;
run;
```

TABLE 3.5a

Solution for Fixed Effects from the MIXED Model for Case 1

Effect	Slide	Rater	Estimate	Standard Error	DF	t Value	Pr < |t|
Intercept			103.49	8.7454	14	11.83	<0.0001
Rater	1		4.2222	4.3641	14	0.97	0.3497
Rater	2	0	.	.	.	.	

SAS output:

Case 1: Replicates are exchangeable (see Table 3.5a)

Covariance Parameter Estimates

Cov Parm	Group	Estimate
Slide*rater		129.02
Slide	rater 1	21.0444
Slide	rater 2	61.8444

$\hat{\mu}_1 - \hat{\mu}_2 = 4.222$, $\hat{\tau}^2 = 129.02$, $\hat{\sigma}_1^2 = 21.0444$, and $\hat{\sigma}_2^2 = 61.8444$. Therefore the LOA are (−32.70, 41.14).

Case 2: Replicates are not exchangeable (see Table 3.5b)

Covariance Parameter Estimates

Cov Parm	Group	Estimate
Slide*rater		129.59
Slide*repl		1.6889
Slide	rater 1	19.3556
Slide	rater 2	60.1556

TABLE 3.5b

Solution for Fixed Effects from the MIXED model for Case 2

Effect	Slide	Rater	Estimate	Standard Error	DF	t Value	Pr > |t|
Intercept		103.49		8.7776	14	11.79	<0.0001
Rater	1		4.22	4.3641	14	0.97	0.3497
Rater	2	0	.	.	.	.	

In this case, we have slight differences in the estimates of the variance components. $\hat{\mu}_1 - \hat{\mu}_2 = 4.222, \hat{\tau}^2 = 129.59, \hat{\sigma}_1^2 = 19.3556, \hat{\sigma}_2^2 = 60.1556$. Hence, the LOA are $(-32.58, 41.02)$.

Note that when we ignore the nature of the replication, we get slightly wider LOA. However, if we use the average of the replications per rater, we definitely get much narrower LOA. This should be expected, since the variance of single observation is smaller than the variance of the average of independent replications.

3.8 Discussion

As we have demonstrated in this chapter, linear regression models play an important role in the analysis of agreement studies. Stockel et al. (1998) argued convincingly that the real life data influence the validity of the regression models being used. Specifically, they pointed out that the quality of the data should be the main concern. Based on their work, Westgard (1998) provided several points of care to consider when analyzing, and interpreting method-comparison data. To summarize, the investigator should be certain that the relationship between the readings made by the two methods is linear. The magnitude of the relative bias and the total variance of the two methods affect the level of agreement (see the expression of concordance correlation). We must be sure that the sampling range is adequately covered. Finally, the study should include adequate number of subjects.

EXERCISES

E3.1 Agreement between two methods in determining AVA. To assess the accuracy of DE in determining AVA, 40 patients were evaluated. AVA was calculated by two methods; the first method is velocity integral and the second method is the maximum velocity (see Table 3.6).

This example analyzes the agreement between the two methods using BAA.

E3.2 Establish the LOA using BAA for the data in Example 3.4, using the average of the replications per slide for each rater. Compare between the LOA and the LOA obtained by fitting the linear mixed model.

E3.3 For the data in Exercise 3.2, fit the Deming regression with proportional errors using the weights: (a) $w_i = 1/x_i^2$, and (b) $w_i = ((x_i + y_i)/2)^{-2}$. Compare the results with the ordinary least-squares fit. Comment on your findings.

E3.4 Consider the simple linear regression model $y_i = \alpha + \beta x_i + e_i$, with $X_i = x_i + u_i$. Furthermore, assume that the elements of the random vector (x_i, e_i, u_i) are independently and normally distributed

TABLE 3.6

Aortic Valve Area Data Measured by Velocity Integral
(Method 1) and Maximum Velocity (Method 2)

Patient	Method 1	Method 2
1	0.8	0.8
2	0.7	0.7
3	1.0	1.0
4	0.8	0.9
5	0.9	0.9
6	0.7	0.8
7	1.0	1.1
8	0.8	0.7
9	0.9	0.9
10	0.6	0.7
11	0.7	0.5
12	1.0	1.0
13	0.7	0.7
14	0.6	0.7
15	0.7	0.7
16	0.6	0.7
17	0.4	0.4
18	0.5	0.6
19	0.4	0.4
20	0.6	0.6
21	0.3	0.3
22	0.5	0.5
23	0.3	0.3
24	0.7	0.7
25	0.6	0.5
26	0.6	0.6
27	0.8	0.9
28	0.7	0.8
29	0.6	0.6
30	0.5	0.4
31	0.3	0.3
32	0.8	0.8
33	0.8	0.7
34	0.5	0.4
35	0.5	0.4
36	0.8	0.7
37	0.6	0.6
38	1.0	1.1
39	0.9	0.8
40	1.0	1.0

random variables with mean vector $(\mu_x, 0, 0)$ and a diagonal variance matrix where the diagonal elements are $(\sigma_x^2, \sigma_e^2, \sigma_u^2)$. Define

$$D_i = y_i - E(y_i) - \gamma(X_i - E(X_i)) \text{ with } \gamma = \text{cov}(y, X)/\sigma_x^2.$$

Show that

1. $Var(D_i) = \sigma_e^2 + \gamma^2\sigma_u^2 + (\beta - \gamma)^2\sigma_x^2.$
2. What is the meaning of σ_u^2 being zero? If σ_u^2 is actually zero, what would be appropriate interpretation of (D_i)?

4

Population Coefficient of Variation as a Measure of Precision and Reproducibility

4.1 Introduction

Suppose that the random variable Y follows some distribution with mean μ and standard deviation σ. The coefficient of variation (CV), denoted by C is defined to be the ratio of the standard deviation to the population mean. That is,

$$C = \frac{\sigma}{\mu}. \tag{4.1}$$

The CV has been found useful in many medical studies. For example, Schwartz et al. (2000) used C to evaluate the repeatability in bidimensional computed tomography measurements of three techniques: hand-held calipers on film, electronic calipers on a workstation, and an auto-counter technique on a workstation. Differences between the coefficients of variation were statistically significant for the auto-counter technique, compared to the other techniques. The CV is often used to compare the dispersion of populations of different scales. For example, in social sciences, when the intent is to compare the variability in school performance with the variability of household income, a comparison of σ's makes no sense because income and school performance are measured on different scales. Another example given by Tian (2005) is from nutritional sciences. In a diet study, the intent was to compare the variability of total/HDL cholesterol with the variability in vessel diameter change. Again, comparison between standard deviations made no sense, because cholesterol and vessel diameter are measured on different scales. In this case, the comparison between the CVs is sensible since scale difference has been adjusted.

In this chapter, we examine the applications and inference procedures on the population CV from single and multiple samples. As a measure of reproducibility, the maximum likelihood estimates (MLE) of the within subject CV is derived and the issues of optimal sampling strategies are discussed.

4.2 Inference from Single Normal Sample

Let $y_1, y_2, \ldots, y_n$ be a simple random sample from a normal population with mean μ and variance σ^2. The MLE of $c = \sigma/\mu$ is given by

$$\hat{c} = \frac{s}{\bar{y}}, \tag{4.2}$$

where $\bar{y} = n^{-1}(y_1 + y_2 + \cdots + y_n)$ and $s = [\sum_{i=1}^{n}(y_1 - \bar{y})^2/n]^{1/2}$.

From Kendall and Ord (1989), it can be shown, to the first order of approximation using the delta method, that

$$\text{var}(c) = \frac{c^2}{2n}(1 + 2c^2). \tag{4.3}$$

An exact confidence interval based on the noncentral t distribution was constructed by Lehmann (1996), but it is computationally intractable. Therefore, several approximate methods have been proposed.

Vangel (1996) proposed approximate methods based on the analysis of the distributions of the class of approximate pivotal quantities for the normal CV. Specifically, an approximate $(1 - \alpha)100\%$ confidence interval based on this approach is given as

$$\left(\frac{\hat{c}}{\sqrt{\ell_1\left(\theta_1\hat{c}^2 + 1\right) - \hat{c}^2}} , \frac{\hat{c}}{\sqrt{\ell_2\left(\theta_2\hat{c}^2 + 1\right) - \hat{c}^2}} \right) \tag{4.4}$$

where

$$\ell_1 = \frac{\chi^2_{v,1-\alpha/2}}{v}, \; \ell_2 = \frac{\chi^2_{v,\alpha/2}}{v}, \; v = n - 1, \; \theta_i = \frac{2}{n\ell_i} + \frac{n-1}{n}, \; i = 1, 2 \text{ and } \chi^2_{v,\alpha}$$

denotes the 100αth percentile of a chi-square random variable. One problem with this interval is the possibility of its nonexistence. This may happen if

$$\ell_i\left(\theta_i\hat{c}^2 + 1\right) \le \hat{c}^2$$

in which case the normal distribution may not be appropriate, and one may use another distribution. For any arbitrary population, the Kendall and Ord

(1989), derived the first-order approximation of the variance of the CV C. Let $\mu_r = E[(y - \mu)^r]$, where $\mu = E(y)$, $\mu_0 = 1$, and $r = 0, 1, 2, 3$.

$$\text{var}(\hat{c}) = \frac{c^2}{n}\left[\frac{\mu_4 - \mu_2^2}{4\mu_2^2} + \frac{\mu_2}{\mu^2} - \frac{\mu_3}{\mu_2\mu}\right]. \tag{4.5}$$

Note that, for the normal case $\mu_3 = 0$, and $\mu_4 = 3\mu_2^2$, and in this case Equation 4.5 reduces to Equation 4.3.

Under the normal distribution setup, one may construct an approximate $(1 - \alpha)100\%$ confidence interval on c, using a variance stabilizing transformation (VST).

Let $g(c)$ be the monotone, differentiable function of c. The conventional choice is that $g(c)$ is chosen so that var$(g)x$ is constant. Since from the delta method

$$\text{var}(g) = \text{var}(c)\left(\frac{\partial g}{\partial c}\right)^2, \tag{4.6}$$

we select

$$g(c)\int \frac{dc}{c\sqrt{1 + 2c^2}}. \tag{4.7}$$

Upon integration we get

$$g(c) = \ell n\left[(1 + 2c^2)^{1/2} - 1\right] - \ell n\left[(1 + 2c^2)^{1/2} + 1\right].$$

Under the above transformation, one may assume that as $n \to \infty$

$$g(\hat{c}) \sim N\big(g(c), \tau_n\big)$$

where $\tau_n = 2/n$. Therefore, the approximate $(1 - \alpha)100\%$ CI on c is

$$CI : (C_\ell, C_u)$$

where

$$C_\ell = \frac{1}{\sqrt{2}}\left[(a_1 - 1)^2 - 1\right]^{1/2},$$

$$C_u = \frac{1}{\sqrt{2}}\left[(a_2 - 1)^2 - 1\right]^{1/2},$$

$$a_i = 2[(1 - \exp(A_i)]^{-1}, \quad i = 1, 2$$

$$A_1 = g(\hat{c}) - z_{\alpha/2}\sqrt{\frac{2}{n}} \qquad (4.8a)$$

$$A_2 = g(\hat{c}) + z_{\alpha/2}\sqrt{\frac{2}{n}}. \qquad (4.8b)$$

4.3 Coefficient of Variation from Gamma Distribution

Let the random variable Y be continuously distributed with PDF of γ with parameters (α, β). The PDF is given by

$$f(y) = \lambda^{\alpha} \ \Gamma^{-1}(\alpha) \ y^{\alpha-1} \ e^{-\lambda y} \quad y, \lambda > 0$$

The rth noncentral moment is given by

$$\mu'_r = \frac{\Gamma(r+\alpha)}{\lambda^r \Gamma(\alpha)},$$

$$\mu = \mu'_1 = \frac{\alpha}{\lambda}, \quad \mu'_2 = \frac{\alpha(\alpha+1)}{\lambda^2},$$

$$\mu'_3 = \frac{\alpha(\alpha+1)(\alpha+2)}{\lambda^3}, \quad \mu'_4 = \frac{\alpha(\alpha+1)(\alpha+2)(\alpha+3)}{\lambda^4},$$

$$\mathrm{var}(y) \equiv \mu_2 = \mu'_2 - \mu^2 = \frac{\alpha}{\lambda^2}.$$

Therefore, the CV is $c = \alpha^{-1/2}$.

To derive the first-order approximation of $\mathrm{var}(\hat{c})$ is given by Equation 4.5, we need the third and fourth central moments of Y:

$$\mu_r = E[Y-\mu]^r, \quad r = 3, 4$$

Hence,

$$\mu_3 = \frac{2\alpha}{\lambda^3}, \quad \text{and} \quad \mu_4 = \frac{3\alpha(\alpha+2)}{\lambda^4}.$$

Therefore,

$$\mathrm{var}(c) = \frac{c^2(1+c^2)}{2n}.$$

4.4 Tests for Equality of Coefficients of Variation

There are several parametric tests for equality of CV from normal populations such as the likelihood ratio test (LRT), the Wald test (WT), and the Score test.

Gupta and Ma (1996) conducted a simulation study to compare the power of these tests and concluded that the LRT has the highest power even when the sample size is small. Another test statistic was introduced by Miller (1991a, 1991b) whose properties were investigated by Feltz and Miller (1996). They called this test DAD. Fund and Tsong (1998) compared these tests and one nonparametric test with respect to their powers. They found that the non-parametric test is robust against departure from the normality assumption.

4.4.1 Likelihood Ratio Test

Let $Y_{i1}, Y_{i2}, \ldots, Y_{ini}$ ($i = 1, 2, \ldots, k$) be k independent random samples of sizes n_i, with $E(Y_{ij}) = \mu_i$ and $\text{var}(Y_{ij}) = \sigma_i^2$, so that $c_i = \sigma_i / \mu_i$ is the ith population CV. We would like to construct a test statistic on the null hypothesis $H_0 : c_i = c$ where c is unknown, against $H_1 : c_i \neq c_j$ for some i, j. Define $\bar{Y}_i = \sum_{j=1}^{n_i} Y_{ij}/n_i$,

$$
S_i = \left[\sum_{j=1}^{n_i} \frac{\left(Y_{ij} - \bar{Y}_i \right)^2}{(n_i - 1)} \right]^{1/2}
$$

Therefore, $\hat{c}_i = S_i/\bar{Y}_i$ is the sample CV.

Under the assumption of normality, the likelihood function under the null hypothesis is proportional to

$$
L_0 = \bar{c}^N \left(\prod_{i=1}^{k} \bar{\mu}_i^{n_i} \right) \exp\left[-\sum_{i=1}^{k} \sum_{j=1}^{n_i} \frac{\left(y_{ij} - \mu_i \right)^2}{2\mu_i^2 c^2} \right] \tag{4.9}
$$

$$
N = \sum_{i=1}^{k} n_i
$$

The likelihood ratio statistic is

$$
M = \sum_{i=1}^{k} n_i \ell n \left(\frac{\tilde{\mu}_i^2 \tilde{c}^2}{S_i^2} \right)
$$

where $\tilde{c}$ and $\tilde{\mu}_i$ are the roots of the equations

$$
-\frac{n_i}{\mu_i} + \sum_{j=1}^{n_i} \frac{y_{ij} \left(y_{ij} - \mu_i \right)}{\mu_i^3 c^2} = 0
$$

$$-\frac{N}{c} + \sum_{i=1}^{k}\sum_{j=1}^{n_i} \frac{\left(y_{ij} - \mu_i\right)^2}{\mu_i^3 c^3} = 0.$$

The asymptotic distribution of M is chi-square with $k - 1$ degrees of freedom under H_0.

4.4.2 Modified Miller's Test

Feltz and Miller proposed the so-called DAD which is given as

$$FM = \bar{c}^2\left[0.5 + c^2\right]^{-1}\left[\sum_{i=1}^{k} v_i\left(c_i - \hat{c}\right)^2\right]^2 \tag{4.10}$$

where $v_i = n_i - 1$, and $\hat{c} = \sum_{i=1}^{k} v_i\hat{c}_i / \sum_{i=1}^{k} v_i$.

The statistic FM is asymptotically chi-square distribution with $k - 1$ degrees of freedom, under H_0. The statistic can be practically used if we replace c by $\bar{C}$.

4.4.3 Nonparametric Test

Let $x_{ij} = Y_{ij}/\mu_i$ for $i = 1, 2, \ldots, k$; $j = 1, 2, \ldots, n_i$. Hence, $E(x_{ij}) = 1$, and $\text{var}\left(x_{ij}\right) = c_i^2$, and testing the equality of CV of the original data is now equivalent to testing the equality of the variances of the scaled data. Usually, μ_i is unknown and must be replaced by $\bar{Y}_i$. Let R_{ij} be the rank of $\left|x_{ij} - 1\right|$. The proposed test statistic is given by

$$SRT = \frac{1}{D^2}\left[\sum_{i=1}^{k} \frac{R_i^2}{n_i} - N\bar{R}^2\right]$$

where $R_i = \sum_{j=1}^{n_i} R_{ij}$, $\bar{R} = 1/N \sum_{i=1}^{k} R_i = (N(2N + 1)/6)$, and $D^2 = N(N+1)(2N+1)$ $(8N + 11)/180$.

Again, the asymptotic distribution of SRT is chi-square with $k - 1$ degrees of freedom.

Example 4.1

The following data represent the Hb measurements in two independent samples of high-school boys and girls from Egypt (Table 4.1).

We shall analyze the data using the above tests statistics. First, for boys $\hat{c}_1 = 4.1(0.92)$, and the 95% confidence interval is (2.26,5.94), and for girls $\hat{c}_2 = 6.9(1.55)$, and the 95% confidence interval is (3.8,10.10). To test the

TABLE 4.1

Hb Measurements for a Sample of 10 Boys and 10 Girls

Boys	12.8	12.7	12.7	13.4	12.5
	12.6	12.6	12.7	11.5	11.9
Girls	12.6	12.4	12.3	13.2	13.1
	11.6	11.4	11.0	11.0	11.9

significance of the difference between the coefficients of variations, we found that the FM statistics is 3.302, and the M statistics is 3.24. Both tests have chi-square with one degree of freedom. Therefore, we conclude that there is no significant difference. For the SRT statistic; $R_1 = 382.5$, $R_2 = 384.4$, and $D^2 = 16,359$. Therefore, SRT = 0.014, supporting the evidence that there is no significant difference.

Sample size issue

$$c = \frac{\sigma}{\mu},$$

$$\mathrm{var}(c) = \frac{c^2}{2n}\left[1 + 2c^2\right],$$

$(1-\alpha)100\%$ CI is

$$c \pm z_{\alpha/2}\frac{c\left(1 + 2c^2\right)^{1/2}}{\sqrt{2n}},$$

$$w = \frac{2z_{\alpha/2}c\left(1 + 2c^2\right)^{1/2}}{\sqrt{2n}} = \frac{cz_{\alpha/2}}{\sqrt{n}}\sqrt{2}\left(1 + 2c^2\right)^{1/2},$$

$$\sqrt{n} = \frac{cz_{\alpha/2}\sqrt{2}\left(1 + 2c^2\right)^{1/2}}{w},$$

$$n = \frac{2c^2 z_{\alpha/2}^2\left(1 + 2c^2\right)}{w^2}.$$

4.5 Statistical Inference on Coefficient of Variation under One-Way Random Effects Model

Under the one-way random effect, we define the CV as

$$c = \frac{\left(\sigma_e^2 + \sigma_b^2\right)^{1/2}}{\mu}.$$

Since c involves σ_b^2 in the numerator, it would receive the same criticism directed at the ICC (Muller and Butter, 1994), in that it is population based. That is, the more is the heterogeneity among subjects (σ_b^2 increases), the larger is the CV. For this reason, we focus in this section on the within-subject coefficient of variation (WSCV). Quan and Shih (1996) defined the WSCV as

$$\theta = \frac{\sigma_e}{\mu}. \tag{4.11}$$

4.6 Maximum Likelihood Estimation

Consider a random sample of k subjects with n repeated measurements of a continuous variable Y, and denote by Y_{ij} the jth reading made on the ith subject under identical experimental conditions ($i = 1, 2, \ldots, k; j = 1, 2, \ldots, n$). In a test–retest scenario, and under the assumption of no reader's effect (i.e., the readings within a specific subject are exchangeable), Y_{ij} denotes the reading of the jth trial made on the ith subject. A useful model for analyzing such data is given by

$$Y_{ij} = \mu + b_i + e_{ij} \quad i = 1,2,\ldots,k; \; j = 1,2,\ldots,n \tag{4.12}$$

where μ is the mean of Y_{ij}, the random subject effects b_i are normally distributed with the mean 0 and variance σ_b^2 or $N(0, \sigma_b^2)$, the measurement errors e_{ij} are $N(0, \sigma_e^2)$, and the b_i and e_{ij} terms are independent. We assume that the subjects are randomly drawn from some population of interest.

Quan and Shih (1996) defined the WSCV parameter in the above model as

$$\theta = \frac{\sigma_e}{\mu}.$$

With model (Equation 4.1), it is assumed that the within-subject variance is the same for all subjects.

Under the above setup, the log-likelihood function has the form

$$l = -\frac{nk}{2}\log 2\pi - \frac{1}{2}k(n-1)\log \sigma_e^2 - \frac{1}{2}k\left[\log\left(\frac{\sigma_e^2}{n+\sigma_b^2}\right)\right]$$

$$-\frac{\sum_i \sum_j (y_{ij}-\mu)^2}{2\sigma_e^2} + \frac{n^2\sigma_b^2\sum_i(\bar{y}_i-\mu)^2}{2\sigma_e^2(\sigma_e^2+n\sigma_b^2)}. \tag{4.13}$$

Define $\sigma^2 = \sigma_b^2 + \sigma_e^2$ and $\rho = \sigma_b^2/(\sigma_b^2 + \sigma_e^2)$, the ICC, for which $\sigma_b^2 = \rho\sigma^2$ and $\sigma_e^2 = \sigma^2(1-\rho)$. Because it is assumed that the design is balanced, the MLE for μ, σ^2, and ρ are given in closed forms by

$$\hat{\mu} = k^{-1}\sum_{i=1}^{k}\bar{y}_i, \quad \hat{\theta} = \frac{\sqrt{MSW}}{\hat{\mu}},$$

$$\sigma_b^2 = \frac{(k-1)MSB - (k)MSW}{nk},$$

and the estimated ICC is

$$\hat{\rho} = \frac{(k-1)MSB - (k)MSW}{(k-1)MSB + k(n-1)MSW},$$

where

$$MSW = \frac{1}{k(n-1)}\sum_{i}^{k}\sum_{j}^{n}(y_{ij} - \bar{y}_i)^2.$$

$MSB = (n/k-1)\sum_{i}^{k}(\bar{y}_i - \hat{\mu})^2$ are, respectively, the within-subject and between-subjects mean-squares as obtained from the usual one-way ANOVA table, and $\bar{y}_i = 1/n\sum_{j=1}^{n}y_{ij}$. Note that the MSB does not exist for $k=1$, which means that to obtain a sensible estimate of ρ as an index of reliability, the study should include more than one subject.

The asymptotic variance–covariance matrix of the MLEs is obtained by inverting Fisher's information matrix. The large sample variance of $\hat{\theta}$ can be obtained using delta method and was shown by Quan and Shih (1996) to be

$$\text{var}(\hat{\theta}) = \frac{A(\rho,n,\theta)}{k} = \frac{\theta^4}{nk}\left[1 + n\frac{\rho}{1-\rho}\right] + \frac{\theta^2}{2k(n-1)}. \tag{4.14}$$

To construct an approximate confidence interval on $\hat{\theta}$, it is assumed that for large k, $\sqrt{k}(\hat{\theta} - \theta)$ follows a normal distribution with mean 0 and variance $A(\rho, b, \theta)$. An approximate $100(1-\alpha)\%$ confidence interval on θ can be given as $\hat{\theta} \pm Z_{\alpha/2}\sqrt{\text{var}(\hat{\theta})}$, where $Z_{\alpha/2}$ is the $100(1-\alpha/2)\%$ cut-off point of the standard normal distribution.

Due to the dependence of the variance of $\hat{\theta}$ on the true parameter value θ itself, Shoukri et al. (2006) found that the asymptotic coverage deviates from its nominal levels for some values of θ. To improve the coverage probability, we suggest a VST to remove the dependence of $\text{var}(\hat{\theta})$ on θ.

4.7 Variance Stabilizing Transformation

To improve the estimated coverage proportion, we use the well-known VST g where $g = \int (\text{var}(\hat{\theta}))^{-1/2} d\theta$. With θ defined as in Equation 4.11, it can be shown that

$$g(\theta) = \sqrt{k(n-1)/2} \, \log \left[\frac{\left(1 + c\theta^2\right)^{1/2} - 1}{\left(1 + c\theta^2\right)^{1/2} + 1} \right]$$

where,

$$c = 2\left(1 - \frac{1}{n}\right)(1 + n\rho^*), \quad \rho^* = \frac{\rho}{(1-\rho)}.$$

Letting

$$f(\theta, n, \rho) = \sqrt{(n-1)/2} \, \log \left[\frac{\left(1 + c\theta^2\right)^{1/2} - 1}{\left(1 + c\theta^2\right)^{1/2} + 1} \right],$$

we may establish assuming the function g is bounded and differentiable, that $f(\hat{\theta}, n, \rho)$ is asymptotically normally distributed with mean $f(\theta, n, \rho)$ and variance $1/k$. Therefore, we can construct $100(1 - \alpha)\%$ confidence limits on θ based on the above transformation. The upper and lower $(1 - \alpha/2)\,100\%$ confidence bound on θ is respectively given by

$$\hat{\theta}_u = \frac{2 \exp\left(\xi_1 \left(2(n-1)\right)^{-1/2} \right)}{c^{1/2} \left[1 - \exp\left(\xi_1 \left(\frac{2}{n-1} \right)^{1/2} \right) \right]}$$

and,

$$\hat{\theta}_i = \frac{2 \exp\left(\xi_2 \left(2(n-1)\right)^{-1/2} \right)}{c^{1/2} \left[1 - \exp\left(\xi_2 \left(\frac{2}{n-1} \right)^{1/2} \right) \right]}$$

where $\xi_1 = f(\hat{\theta}, n, \rho) + z_{\alpha/2}/\sqrt{k}$, and $\xi_2 = f(\hat{\theta}, n, \rho) - z_{\alpha/2}/\sqrt{k}$.

Example 4.2

Accurate and reproducible quantification of brain lesion count and volume in MS patients using MRI is a vital tool for the evaluation of disease progression and patient response to therapy. Current standard methods for obtaining these data are largely manual and subjective and are therefore, error-prone and subject to inter- and intra-operator variability. Therefore, there is a need for a rapid automated lesion quantification method. Ashton et al. (2003) compared manual measurements and an automated data technique known as GEORG of the brain lesion volume of three MS patients, each measured 10 times by a single operator for each method. The data are presented in Table 4.2.

Based on the guidelines for the levels of reliability provided by Fleiss (1986), a value of an ICC above 80% indicates an excellent reliability, and from Table 4.3 both methods cross this threshold level. However, based on the WSCV values, the manual method is definitely less reproducible than the automated method (the GEORG is five times more reproducible than the manual method). This example demonstrates the usefulness of the WSCV over the ICC as a measure of reproducibility. Clearly, one should construct a formal test on the significance of the difference between two correlated within-subject CV. There are several competing methods to construct such a test (e.g., LRT, WT, and Score test) and we shall discuss this issue in the last section of this chapter.

TABLE 4.2

Results for 10 Replicates on Each of Three Patient's Total Lesion Burden

Patient	Method[a]	\multicolumn{10}{c}{Replicates}

Patient	Method[a]	1	2	3	4	5	6	7	8	9	10
1	M	20	21.2	20.8	20.6	20.2	19.1	21	20.4	19.2	19.2
	G	19.5	19.5	19.6	19.7	19.3	19.1	19.1	19.3	19.2	19.5
2	M	26.8	26.5	22.5	23.1	24.3	24.1	26	26.8	24.9	27.7
	G	22.1	21.9	22	22.1	21.9	21.8	21.7	21.7	21.7	21.8
3	M	9.6	10.5	10.6	9.2	10.4	10.4	10.1	8	10.1	8.9
	G	8.5	8.5	8.3	8.3	8.3	8	8	8	8	8.1

Note: Values are given volumes in cubic centimeters.
[a] M = manual, G = geometrically constrained region growth.

TABLE 4.3

Summary Analysis of Data in Table 4.2 and 95% Confidence Intervals

Method	$\hat{\rho}$	$\hat{\theta}$	95% CI without VST	95% CI with VST
M	0.966	6.5%	(0.034, 0.096)	(0.043, 0.118)
G	0.999	1.2%	(0.006, 0.017)	(0.008, 0.021)

4.8 Estimating WSCV when Error Variance Is Not Common

Note that under the one-way random effects model, computation of the WSCV is based on the assumption that the intrasubject variation remains unchanged from subject to subject. However, the assumption that the intra-subject variation is constant may not be satisfied. Several methods to estimate the overall CVs have been developed. To demonstrate how this situation is dealt with we assume, under the one-way random effects model, that $\text{var}(y_{ij}) = \sigma_s^2 + \sigma_{ei}^2$ $i = 1, 2, \ldots, k$.

Hence, we define the within-subject CV as $\theta_i = \sigma_{ei}/\mu$.

Let $\bar{y}_i$ and s_i^2 be the mean and the variance for the ith subject, where

$$\bar{y}_i = \frac{1}{n}\sum_{j=1}^{n} y_{ij}, \quad s_i^2 = \frac{1}{n-1}\sum_{j=1}^{n}\left(y_{ij} - \bar{y}_i\right)^2.$$

An intuitive approach is to estimate θ_i by $s_i/\bar{y}_i$. Several estimators were proposed by Chow and Tse (1990). The first one is the given as

$$\hat{\theta}_1 = \frac{1}{k}\sum_{i=1}^{k}\frac{s_i}{\bar{y}_i}.$$

The second one is given as

$$\hat{\theta}_2 = \left\{\frac{1}{k}\sum_{i=1}^{k}\left(\frac{s_i}{\bar{y}_i}\right)^2\right\}^{1/2}.$$

In addition to the above estimators, Chow and Tse (1990) suggested

$$\hat{\theta}_3 = \sum_{i=1}^{k} s_i\bar{y}_i \bigg/ \sum_{i=1}^{k}\left(\bar{y}_i\right)^2,$$

$$\hat{\theta}_4 = \left\{\sum_{i=1}^{k} s_i^2\bar{y}_i^2 \bigg/ \sum_{i=1}^{k}\bar{y}_i^4\right\}^{1/2},$$

and

$$\hat{\theta}_5 = \left\{\frac{MSW}{\bar{y}^2 + \hat{\sigma}_s^2}\right\}^{1/2}.$$

where

$$\hat{\sigma}_s^2 = \frac{MSB - MSW}{n}, \quad \text{and} \quad \bar{y} = \sum_{i=1}^{k} \frac{\bar{y}_i}{k}.$$

Also, the definition of the proposed estimators will not be affected if the number of replicates n_i ($i = 1, 2, ..., k$) is not the same.

Example 4.3

As an example, we provide hypothetical data representing the area under the curve (AUC) of one drug formulation, for 15 subjects. Each subject received three administration of the given drug formulation. The data are given in Table 4.4.

A regression of log (s_i) and $\log(\bar{y}_i)$ shows a significant relationship between $\bar{y}_i$ and s_i, such that

$$\log(s_i) = -15.4 + 2.53\left[\log(\bar{y}_i)\right].$$

The value of the *F*-statistic for testing the null hypothesis that the regression coefficient is zero is $F = 7.94$ (*P*-value = 0.0145).

Note that the *WSCV* obtained from the one-way random effects model (assuming constant within-subject variance component) is given by $\hat{\theta} = 18.31$.

TABLE 4.4

Area under the Curve of an Intravenous Formulation

Subject	AUC (1)	(2)	(3)	$\bar{y}_i$	s_i	cv_i
1	6835	7000	8907	7580.7	1151.59	15.19
2	6523	8041	6236	6933.3	969.94	13.99
3	8644	6153	8801	7866.0	1485.58	18.88
4	9168	6757	6192	7372.3	1580.54	21.44
5	790	6768	9124	7930.7	1178.30	14.86
6	8178	6330	9211	7906.3	1459.59	18.46
7	6534	6950	9670	7718.0	1703.23	22.07
8	5878	8066	6512	6818.7	1125.78	16.51
9	9552	6551	8988	8363.7	1594.94	19.07
10	6186	6043	8030	6753.0	1108.22	16.41
11	5694	7140	8600	7144.7	1453.01	20.34
12	7463	6213	5784	6486.7	872.3	13.45
13	7535	5338	7372	6748.3	1224.1	18.14
14	6509	10100	6666	7758.3	2029.5	26.16
15	7530	6400	7100	7010.0	570.35	8.14

The four estimates proposed by Chow and Tse (1990) are

$$\hat{\theta}_1 = 17.5, \quad \hat{\theta}_2 = 18, \quad \hat{\theta}_3 = 17.79, \quad \hat{\theta}_4 = 18.3.$$

Chow and Tse noted that, since $MSB < MSW$, then $\hat{\sigma}_s^2$ is negative (Thompson, 1962), and for the case $MSB < (k + 1/k - 1)MSW$, the proposed estimates for $\hat{\sigma}_s^2$ is

$$\hat{\sigma}_s^2 = \frac{k - 1}{nk(k + 1)} MSB.$$

The ANOVA results, under the one-way random effect specifications, give $MSB = 936953$, $MSW = 1816384$. Therefore, $\hat{\sigma}_s^2 = 18218.5$, and from which we have $\hat{\theta}_5 = 18.3$.

As can be seen, the five estimates do not substantially differ from the WSCV estimator. However, investigating the statistical properties of these estimators in terms of bias and precision for a wide spectrum of parameter values, and sample sizes might reveal which estimator is best.

4.9 Sample Size Estimation

In the following development, we investigate the issue of sample size estimation. We assume that the investigator is interested in the number of replicates, n, per subject, so that the variance of the estimate of θ is minimized, given that the total number of measurements is fixed *a priori* at $N = nk$.

4.9.1 Efficiency Criterion

For fixed total number of measurements $N = nk$, Equation 4.14 gives

$$\mathrm{var}\left(\hat{\theta}\right) = \frac{\theta^4}{N}\left[1 + n\frac{\rho}{1 - \rho}\right] + \frac{n\theta^2}{2N(n - 1)}. \tag{4.15}$$

The necessary condition for $\mathrm{var}(\hat{\theta})$ to have a unique minimum is that $\partial\mathrm{var}(\hat{\theta})/\partial n = 0$. This, and the additional condition that $\partial^2\mathrm{var}(\hat{\theta})/\partial n^2 > 0$ are both satisfied as long as $0 < \rho < 1$. Differentiating (Equation 4.15) with respect to n, equating to zero and solving for n we obtain

$$n^* = 1 + \sqrt{\frac{(1 - \rho)}{2\rho\theta^2}} \quad 0 < \rho < 1. \tag{4.16}$$

Table 4.5 gives a few allocation schemes of (n,k) for $\rho = 0.6$, 0.7, and 0.8, $\theta = 0.1$, 0.2, 0.3, and 0.4, when $N = 24$. $k^* = N/n^*$.

TABLE 4.5

Optimal Combinations of (n_{opt}, k_{opt}), which Minimize the Variance of $\hat{\theta}$ for $N = 24$

θ	0.6	0.7	0.8
		(n_{opt}, k_{opt})	
0.1	(6.77, 3.54)	(5.63, 4.26)	(4.54, 5.29)
0.2	(3.89, 6.17)	(3.31, 7.24)	(2.77, 8.67)
0.3	(2.91, 8.23)	(2.54, 9.47)	(2.17, 11.05)
0.4	(2.44, 9.82)	(2.16, 11.12)	(1.88, 12.74)

Note that in practice, only integer values of (n, k) are used, and because $N = nk$ is fixed *a priori*, we first round the optimum values of n to the nearest integer; then $k = N/n$ was rounded to the nearest integer. It is clear that to efficiently estimate the WSCV for large values of θ, we need a smaller number of replicates and a larger number of subjects.

4.9.2 Fixed with Confidence Interval Approach

Bonett (2002) discussed the issue of sample size requirements that achieve a pre-specified expected width for a confidence interval about ICC. This approach is useful in planning a reliability study in which the focus is on estimation rather than hypothesis testing. He demonstrated that the effect of inaccurate planning value of ICC is more serious in hypothesis testing applications. Shoukri et al. (2003) argued that the hypothesis testing approach might not be appropriate while planning a reproducibility study. This is because, in most cases, values of the coefficient under the null and alternative hypotheses may be difficult to specify. An alternative approach is to focus on the width of the CI for θ. Since the approximate width of an $(1 - \alpha)100\%$ CI on θ is $2z_{\alpha/2}\text{var}(\hat{\theta})^{1/2}$, an approximate sample size that yields an $(1 - \alpha)100\%$ CI for θ with a desired width w obtains by setting $w = 2z_{\alpha/2}\{\text{var}(\hat{\theta})\}^{1/2}$ and then solving for k

$$k = \frac{4z^2 A(\rho, n, \theta)}{w^2}. \tag{4.17}$$

We observe that, for fixed n and θ, larger values of ρ require a larger number of subjects to satisfy the criterion. As an example, suppose that it is of interest to construct 95% CI on θ with expected width $w = 0.05$, $\rho = 0.3$, and an afforded number of replicates $n = 2$. If the hypothesized value of θ is 0.10, then $k = 31$, and if θ is 0.3 (i.e., lower reliability), then $k = 323$.

4.9.3 Cost Criterion

Funding constraints often determine the cost of recruiting subjects for a reliability study. Although too small a sample may lead to a study that produces an imprecise estimate of the reproducibility coefficient, too large a sample may result in a waste of resources. Thus, an important decision in a typical reliability study is to balance the cost of recruiting subjects with the need for a precise estimate of the parameter summarizing reliability.

In this section, we determine the combinations (n, k) that minimize the variance of $\mathrm{var}(\hat{\theta})$ subject to cost constraints. Constructing a flexible cost function starts with identifying sampling and overhead costs. The sampling cost depends primarily on the size of the sample and includes costs for data collection, compensation to volunteers, management, and evaluation. On the other hand, overhead costs are independent of sample size. Following Sukhatme et al. (1984), we assume that the overall cost function is given as

$$C = c_0 + kc_1 + nkc_2 \qquad (4.18)$$

where c_0 is the fixed cost, c_1 the cost of recruiting a single subject, and c_2 is the cost of making one observation. Using the method of Lagrange multipliers and following Shoukri et al. (2003), we write the objective function ψ in the form

$$\psi = \mathrm{var}\left(\hat{\theta}\right) + \lambda\left(C - c_0 - kc_1 - nkc_2\right) \qquad (4.19)$$

where $\hat{\theta}$ is given by Equation 4.14 and λ is the Lagrange multiplier. Differentiating ψ with respect to n, k, and λ and equating to zero, we obtain

$$2\theta^2\rho^* n^4 - 4\theta^2\rho^* n^3 - \left(2\theta^2 r + r - 2\theta^2\rho^* + 1\right)n^2 + 4\theta^2 rn - 2\theta^2 r = 0 \quad (4.20)$$

where $r = c_1/c_2$, and $\rho^* = \rho/(1 - \rho)$.

Although an explicit solution to Equation 4.20 is available, the resulting expression is complicated and does not provide any useful insight. The 4th degree polynomial on the left side of Equation 4.20 has two imaginary roots, one negative and one admissible (positive) root for n. Table 4.6 summarizes the results of the optimization procedure where we provide the optimal n for various values of θ, ρ, and r, noting that

$$k_{opt} = \frac{(C - c_0)/c_2}{r + n_{opt}}. \qquad (4.21)$$

From Table 4.6, it is apparent that when $r = c_1/c_2$ increases, the required number of replicates per subject (n) increases, because the cost of making a single observation (c_2) decreases and the cost of recruiting a subject (c_1) increases. When r is fixed, an increase in ρ, results in a decline in the required value of n, and accordingly an increase in k. An increase in θ, also results in a decrease in n. The general conclusion is that it is sensible to decrease the number of items associated with a higher cost, while increasing those with lower cost.

TABLE 4.6

Optimal Replications (Rounded to the Nearest Integer) of n that Minimize the Variance of $\hat{\theta}$ Subject to Cost Constraints

		r						
		0.1	0.5	1	1.5	5	10	20
θ	ρ				n_{opt}			
0.01	0.6	62	72	83	92	142	193	266
	0.7	50	58	66	74	114	155	213
	0.8	38	44	52	57	88	118	163
0.04	0.6	16	19	21	24	36	49	67
	0.7	13	15	17	19	29	39	54
	0.8	10	12	14	15	23	30	42
0.10	0.6	7	8	9	10	15	20	28
	0.7	6	7	8	8	12	17	22
	0.8	5	5	6	7	10	13	17
0.15	0.6	5	6	7	7	11	14	19
	0.7	4	5	5	6	9	11	15
	0.8	3	4	4	5	7	9	12
0.30	0.6	3	3	4	4	6	8	10
	0.7	3	3	3	4	5	6	9
	0.8	2	2	3	3	4	5	8
0.40	0.6	3	3	3	3	5	6	8
	0.7	2	2	3	3	4	5	7
	0.8	2	2	2	2	3	4	5

We note that by setting $c_1 = 0$ in Equation 4.20, we obtain $n_{opt} = 1 + \sqrt{(1-\rho)/2\rho\theta^2}$, as in Equation 4.16. The situation $c_1 = 0$ is quite plausible, at least approximately if the major cost is in actually making the observations (e.g., expensive equipment, cost of interviews versus free volunteer subjects). This means that a special cost structure is implied by the optimal allocation procedure discussed earlier.

Example 4.4

To assess the accuracy of DE in determining AVA prospective evaluation on patients with AS, an investigator wishes to demonstrate a high degree of reliability ($\rho = 0.80$) in estimating AVA using the "velocity integral method" with a planned value for the WSCV = 0.10. Suppose that the total cost of making the study is fixed at $1600.00. It is assumed that the overhead fixed cost c_0 is absorbed by the hospital. Moreover, we assume that travel cost is $200.00, and the administrative cost using the DE is $200.00 per visit. From Table 4.5, n_{opt} for $r = 1$, $\rho = 0.8$, and $\theta = 0.10$ is 6. From Equation 4.21, $k_{opt} = (1600 / 15) / (1 + 6) = 15$. That is, we need 15 patients with 6 measurements each to minimize var($\hat{\theta}$) subject to the given cost.

4.10 Analysis of WSCF from Two Dependent Samples

Suppose that for the data in Example 4.2, we are interested in comparing the CV of the two methods (M, G). Since we are measuring the same subjects by both methods, the estimated CVs are bound to be correlated.

Shoukri et al. (2008) reported on an important application from molecular biology research in which correlated/dependent reproducibility coefficients are compared in terms of reproducibility of gene expression measurements. DNA microarrays are powerful technologies, which make it possible to study genome-wide gene expressions and are extensively used in biological research. As the technology evolves rapidly a number of different platforms became available, which introduces some challenges for researchers to know which technology is best suited for their needs. There have been various studies that directly compared the performance of one platform with another in terms of cross-platform comparability and agreement of gene expression results. However, the results of these studies are conflicting: some demonstrate concordance, others discordance between technologies. Thus, one needs to take into consideration the accuracy and reproducibility of different types of microarrays when allocating the laboratory resources for future experiments.

The following setup is similar to what we have done when our interest is to compare two dependent reliability coefficients. In the next section, we shall present methodologies for comparing two dependent coefficients of variations.

4.10.1 Likelihood-Based Methodology

Suppose that we are interested in comparing the reproducibility of two instruments. Let x_{ijl}, be the jth measurement of the ith subject by the lth instrument, $j = 1, 2, ..., m_l$, $i = 1, 2, ..., n$, and $l = 1, 2$. To evaluate the WSCV, we consider the one-way random effects model

$$x_{ijl} = \mu_l + b_i + e_{ijl} \qquad (4.22)$$

where μ_l, is the mean value of measurements made by the lth instrument, b_i are independent random subject effects with $b_i \sim N(0, \sigma_b^2)$, and e_{ijl} are independent $N(0, \sigma_l^2)$. Many authors have used the ICC, ρ_l defined by the ratio $\rho_l = \sigma_b^2/(\sigma_b^2 + \sigma_l^2)$.

For the lth sample, we define WSCV as $\theta_l = \sigma_l/\mu_l$ as a measure of reproducibility. It determines the degree of closeness of repeated measurements taken on the same subject either by the same instruments or on different occasions under the same conditions. It is clear that the smaller the WSCV, the better the reproducibility.

The following setup is to facilitate the construction of the likelihood function.

Let

$$X_i = \left(X_{i1},\ X_{i2},\ \ldots,X_{im_1},\ X_{i,m_1+1},\ X_{i,m_1+2},\ldots,\ X_{i,m_1+m_2}\right)'$$

denote the measurements on the ith subject, $i = 1, 2, \ldots, n$ where $X_{i1}, X_{i2}, \ldots,$ X_{im_1} are the m_1 measurements obtained by the first method (platform), $X_{i,m_1+1}, X_{i,m_1+2}, \ldots, X_{i,m_1+m_2}$ the m_2 measurements obtained the second method (platform). We assume that $X_i \sim N(\mu, \Sigma)$, where $\mu^T = (\mu_1 1_{m_1}^T, \mu_2 1_{m_2}^T)$ and,

$$\Sigma = \begin{bmatrix} \sigma_1^2 I_{m_1} + \dfrac{\rho_1}{1-\rho_1}\sigma_1^2 J_{m_1} & \rho_{12}\sigma_1^* \sigma_2^* J_{m_1 x m_2} \\[4mm] \rho_{12}\sigma_1^* \sigma_2^* J_{m_1 x m_2} & \sigma_2^2 I_{m_2} + \dfrac{\rho_2}{1-\rho_2}\sigma_2^2 J_{m_2} \end{bmatrix} \tag{4.23}$$

where $\sigma_\ell^* = \sigma_\ell(1-\rho_\ell)^{-1/2}$, $\ell = 1, 2$.

In these expressions 1_k is a column vector with all k elements equal to 1, I_k is an $k \times k$ identity matrix and J_k and J_{kxt} are $k \times k$ and $k \times t$ matrices with all the elements equal to 1. Thus, the model assumes that the m_1 observations taken by the first method have common mean μ_1, common variance σ_1^2, and common ICC ρ_1, whereas the m_2 measurements taken by the second method have common mean μ_2, common variance σ_2^2, and common ICC ρ_1. Moreover, ρ_1 denotes the interclass correlation between any pair of measurements $x_{ij}(j = 1,2,\ldots,m_1)$ and $x_{im1+t}(t = 1,2,\ldots,m_2)$, and also assumed constant across all subjects in the population.

For the lth method, the WSCV, which will be denoted as θ_l is defined as

$$\theta_l = \frac{\sigma_l}{\mu_l}, \quad l = 1, 2.$$

Our primary aim is to develop and evaluate methods of testing $H_0 : \theta_1 = \theta_2$, taking into account dependencies induced by a positive value ρ_{12}. We restrict our evaluation to reproducibility studies having $m_1 = m_2 = m$.

4.10.2 Methods for Testing Null Hypothesis

4.10.2.1 Wald Test

If $X_1, X_2, \ldots, X_n$ is a sample from the above multivariate normal distribution, then the *log*-likelihood function l, as a function of $\psi = (\mu_1, \mu_2, \sigma_1^2, \sigma_2^2, \rho_1, \rho_2, \rho_{12})$ is given as

$$-2L = Q + nm\log\left(\sigma_1^2\sigma_2^2\right) - n\log\left((1-\rho_1)(1-\rho_2)\right) + n\log w \tag{4.24}$$

where

$$w = u_1 u_2 - m^2 \rho_{12}^2,$$

$$u_l = 1 + (m-1)\rho_l, \quad l = 1, 2$$

and,

$$Q = \frac{S_1^2}{\sigma_1^2} + \frac{m(1-\rho_1)u_2}{w\sigma_1^2} \sum_{i=1}^{n} (\bar{x}_{i1} - \mu_1)^2 + \frac{S_2^2}{\sigma_2^2} + \frac{m(1-\rho_2)u_1}{w\sigma_2^2} \sum_{i=1}^{n} (\bar{x}_{i2} - \mu_2)^2$$

$$- \frac{2m^2\rho_{12}}{w\sigma_1\sigma_2} \left((1-\rho_1)(1-\rho_2) \right)^{1/2} \sum_{i=1}^{n} (\bar{x}_{i1} - \mu_1)(\bar{x}_{i2} - \mu_2).$$

From the conditions $\{1 + (m-1)\rho_1\}\{1 + (m-1)\rho_2\} > m^2\rho_{12}^2$ and $-1/(m-1) < \rho_l < 1$ must be satisfied for the likelihood function to be a sample from a nonsingular multivariate normal distribution.

The summary statistics given in (Q) are defined as

$$\bar{x}_{ij} = \sum_{k=1}^{m} x_{ijk}/m \quad i = 1, 2, \dots, n; \quad j = 1, 2$$

$$S_j^2 = \sum_{i=1}^{n} \sum_{k=1}^{m} \left(x_{ijk} - \bar{x}_{ij} \right)^2.$$

The MLE for μ_l and σ_l^2 are given respectively by $\bar{\mu}_l = \bar{x}_l, \hat{\sigma}_l^2 = S_l^2/n(m-1)$, where $\bar{x}_l = 1/n \sum_{i=1}^{n} \bar{x}_{ij}$ and $l = 1, 2$. Clearly, $\hat{\sigma}_l^2$ exists for values of $m > 1$. Therefore, we shall assume that $m > 1$ throughout this chapter. From Donner and Zou (2002), we obtain $\hat{\rho}_1$ and $\hat{\rho}_2$ by computing Pearson's product–moment correlation over all possible pairs of measurements that can be constructed within methods 1 and 2, respectively, with $\hat{\rho}_{12}$ similarly obtained by computing this correlation over the nm^2 pairs $(x_{ij}, x_{i'm+l})$.

The WT of $H_0 : \theta_1 = \theta_2$ requires the evaluation of variance of $\hat{\theta}_l, l = 1, 2$, and $\mathrm{cov}(\hat{\theta}_1, \hat{\theta}_2)$. To obtain these values we use elements of Fisher's information matrix (see Appendix at the end of this chapter), along with the delta method (see Chapter 1). On writing $\psi = (\psi_1, \psi_2)', \psi_1 = (\mu_1, \mu_2)'$ and $\psi_2 = (\sigma_1^2, \sigma_2^2, \rho_1, \rho_2, \rho_{12})'$, the Fisher's information matrix $I = -E[\partial^2 l/\partial\psi\partial\psi']$ has the following structure:

$$I = \begin{bmatrix} I_{11} & 0 \\ 0 & I_{22} \end{bmatrix} \tag{4.25}$$

Note that, $I_{12} = I_{21} = -E(\partial^2 l / \partial \psi_1 \partial \psi_2) = 0$. Therefore, from the asymptotic theory of maximum likelihood estimation, we have

$$I_{11}^{-1} = \begin{bmatrix} \text{var}(\hat{\mu}_1) & \text{cov}(\hat{\mu}_1, \hat{\mu}_2) \\ \text{cov}(\hat{\mu}_1, \hat{\mu}_2) & \text{var}(\hat{\mu}_2) \end{bmatrix},$$

and the elements of I_{22} are given in the appendix.

The elements of I_{22}^{-1} are the asymptotic variance–covariance matrix of the MLE of the covariance parameters. Inverting Fisher's information matrices, we get

$$\text{var}(\hat{\mu}_l) = \frac{\sigma_l^2}{nm(1-\rho_1)}\left[1 + (m-1)\rho_l\right] \quad l = 1, 2. \tag{4.26}$$

Applying the delta method, we can show, to the first order of approximation that

$$\text{var}(\hat{\sigma}_l) \approx \frac{\sigma_l^2}{2n}(m-1) \quad l = 1, 2. \tag{4.27}$$

The MLE of θ_l is, $\hat{\theta}_l = (\hat{\mu}_l / \hat{\sigma}_l)$. Again, by application of the delta method, we can show to the first order of approximation that

$$\text{var}(\hat{\theta}_l) \approx \frac{\theta_l^4\left[1 + (m-1)\rho_l\right]}{nm(1-\rho_1)} + \frac{\theta_l^2}{2n(m-1)} \tag{4.28}$$

as was shown by Quan and Shih (1996).

Again using the delta method, we show approximately that

$$\text{cov}(\hat{\theta}_1, \hat{\theta}_1) \approx \frac{2\theta_1^2 \theta_2^2 \rho_{12}}{n\sqrt{(1-\rho_1)(1-\rho_2)}}. \tag{4.29}$$

We apply the large sample theory of maximum likelihood to establish that

$$Z = \frac{\hat{\theta}_1 - \hat{\theta}_2}{\sqrt{\text{var}(\hat{\theta}_1) + \text{var}(\hat{\theta}_2) - 2\text{cov}(\hat{\theta}_1, \hat{\theta}_2)}} \tag{4.30}$$

is approximately distributed under H_0 as a standard normal deviate. The denominator of Z is the standard error of $(\hat{\theta}_1 - \hat{\theta}_2)$ and is denoted by $SE(\hat{\theta}_1 - \hat{\theta}_2)$. Since the standard error of $(\hat{\theta}_1 - \hat{\theta}_2)$ contains unknown parameters, its maximum likelihood estimate $SE(\hat{\theta}_1 - \hat{\theta}_2)$ is obtained by sub-stituting $\hat{\theta}_l$ for θ_l, $\tilde{\rho}_l$ for ρ_l and $\tilde{\rho}_{12}$ for ρ_{12}. Moreover, we may construct an approximate $(1 - \alpha)100\%$ confidence interval on $(\theta_1 - \theta_2)$ given as

$$\hat{\theta}_1 - \hat{\theta}_2 \pm z_{\alpha/2}\hat{SE}(\hat{\theta}_1 - \hat{\theta}_2),$$

where $z_{\alpha/2}$ is the $(1 - \alpha/2)100\%$ cut-off point of the standard normal distribution.

4.10.2.2 Likelihood Ratio Test

The following algorithmic approach was given by Shoukri et al. (2008). An LRT of $H_0 : \theta_1 = \theta_2$ was developed numerically, and computed by first setting $\mu_l = \sigma_l/\theta_l$, $l = 1, 2$ in Q, and then adopting the following algorithm:

1. Set $\mu_l = \sigma_l/\theta_l$, $l = 1,2$ in Q, thereafter.
2. Set $\theta_1 = \theta_2 = \theta$ in Q.
3. Minimize the resulting expression with respect to all six parameters $(\sigma_1, \sigma_2, \rho_1, \rho_2, \rho_{12}, \theta_1, \theta_2)$.
4. Subtract the minimum from the minimum of $-2L$ as computed over all seven parameters $(\sigma_1, \sigma_2, \rho_1, \rho_2, \rho_{12}, \theta_1, \theta_2)$ in the model.

It then follows from standard likelihood theory that the resulting test statistic is approximately chi-square distributed with 1 degree of freedom under H_0.

4.10.2.3 Score Test

One of the advantages of the likelihood-based inference procedure is that in addition to the WT and the LRT "Rao's score test" can also be readily devel-oped. The motivation for it is that it can sometimes be easier to maximize the likelihood function under the null hypothesis than under the alternative hypothesis. A standard procedure for performing the score test of $H_0 : \theta_1 = \theta_2$, is to set $\theta_2 = \theta_1 + \Delta$ so that the null hypothesis is equivalent to $H_0 := 0$ where Δ is unrestricted. Replacing μ_l by σ_l/θ_l, the log-likelihood function L is then independent of μ_l.

Let $L = L(\Delta; \psi^*) = L(\Delta; \theta_1, \sigma_1, \sigma_2, \rho_1, \rho_2, \rho_{12})$ and $l_1 = (\partial L/\partial \Delta)$, $l_2 = (\partial L/\partial \psi)$. From Cox and Hinkley (1974), the score statistic is given by

$$S = i_1^T A_{1\bullet2}^{-1} i_1,$$

where

$$l_1^* = \frac{\partial L}{\partial \Delta}\Big|_{\Delta=0} = \frac{nm}{\hat{w}\hat{\theta}_1^2}\left[\hat{\mu}_1\left(1 - \hat{\rho}_1\right)\left(\frac{\hat{\theta}_1 - \hat{\theta}_2}{\hat{\theta}_1\hat{\theta}_2}\right)\right] \tag{4.31}$$

and $A_{1\bullet2} = A_{11} - A_{12}A_{12}^{-1}A_{21}$. The matrices on the right-hand side of $A_{1\bullet2}$ are obtained from partitioning the Fisher's information matrix A so that

$$A = \begin{pmatrix} A_{11} & A_{12} \\ A_{21} & A_{22} \end{pmatrix}$$

where

$$A_{11} = E\left(\frac{-(\partial^2 l)}{(\partial\Delta\partial\psi^*)} \right),$$

$$A_{12} = A_{21}^T = E\left(\frac{-(\partial^2 l)}{(\partial\Delta\partial\psi^*)} \right),$$

and

$$A_{22} = E\left(\frac{-(\partial^2 l)}{(\partial\psi^*\partial\psi^{*T})} \right)$$

with all the matrices on the right-hand side of $A_{1\bullet2}$ evaluated at $\Delta = 0$. When an estimator other than the MLE is used for the nuisance parameters ψ^*, provided that the estimator ψ^* is $\sqrt{n}$ consistent, it was shown that the asymptotic distribution of S is that of a chi-square with 1 degree of freedom (see Neyman, 1959).

The score test has been applied in many situations and has been proven to be locally powerful. Unfortunately, the inversion of $A_{1\bullet2}$ is quite complicated and we cannot obtain a simple expression for S that can be easily used. Moreover, it has been found through extensive simulations that while the score test holds its levels of significance, it is less powerful than LRT and WT across all parameter configurations.

4.10.2.4 Regression Test

We have used in Chapter 3, the Pitman (1939) and Morgan (1939) technique to test the equality of variances of two correlated normally distributed random variables. It is constructed to simply test for zero correlation between the sums and differences of the paired data. Recall from Chapter 3 that Bradley and Blackwood (1989) extended Pitman and Morgan's (PM) idea to a regression context that affords a simultaneous test for both the means and the variances. The test is applicable to many paired data settings, for example, in evaluating the reproducibility of lab test results obtained from two different sources. The test could also be used in repeated measures experiments, such as in comparing the structural effects of two drugs applied to the same set of subjects. Here we generalize the results of Bradley and Blackwood to establish the simultaneous equality of means and variances of two correlated variables, implying the equality of their coefficients of variations.

Let $\bar{X}_{ij} = \sum_{k=1}^{m} X_{ijk}/m$, and define $d_i = \bar{X}_{i1} - \bar{X}_{i2}$, and $s_i = \bar{X}_{i1} + \bar{X}_{i2}$.

Direct application of the multivariate normal theory shows that the conditional expectation of d_i on s_i is linear. That is,

$$E\left(d_i \mid s_i\right) = \alpha + \beta s_i, \tag{4.32}$$

where

$$\alpha = \left(\mu_1 - \mu_2\right) - \left(\mu_1 + \mu_1\right)\left(\sigma_1^2 - \sigma_2^2\right)\zeta \tag{4.33}$$

$$\beta = \left(\sigma_1^2 - \sigma_1^2\right)\zeta \tag{4.34}$$

where

$$\zeta^{-1} = \sigma_1^2(1 - \rho_1)^{-1}(1 + (2m - 1)\rho_1) + \sigma_2^2(1 - \rho_2)^{-1}(1 + (2m - 1)\rho_2)$$

is strictly positive.

The proof is straightforward and is therefore, omitted.

It can be shown then from direct application of the multivariate normal theory that the conditional expectation (Equation 4.32) is linear, and does not depend on the parameter ρ_{12}.

From Equations 4.33 and 4.34 it is clear that $\alpha = \beta = 0$ if and only if $\mu_1 = \mu_2$ and $\sigma_1 = \sigma_2$ simultaneously. Therefore, testing the equality of two correlated CV is equivalent to testing the significance of the regression Equation 4.32. From the theory of least squares, if we define

$$TSS = \sum_{i=1}^{n} 90(d_i - \bar{d})^2, \quad RSS = \hat{\beta}_1^2 \sum_{i=1}^{n} (s_i - \bar{s})^2$$

and

$$EMS = \frac{(TSS - RSS)}{n - 2},$$

the hypothesis $H_0 = \alpha = \beta = 0$ is rejected when RSS/EMS exceeds $F_{v,\,1,(n-2)}$, the $(1 - v)100\%$ percentile value of the F-distribution with 1 and $(n - 2)$ degrees of freedom.

Example 4.5

In this example, we demonstrate the statistical methodologies of this chapter on a small data set. The data are from a study using the computer-aided tomographic scans (CAT-SCAN) of the heads of 50 psychiatric patients (Dunn, 1992). The data

TABLE 4.7

Analysis of Computer-Aided Tomographic Scan Data on 50 Patients
via PIX or PLAN with Two Replicates

	PIX ($l = 1$)		PLAN ($l = 2$)	
	Estimate	SE	Estimate	SE
μ_1	1.41	0.074	1.79	0.056
ρ_1	0.99	0.002	0.73	0.066
θ_1	0.028	0.003	0.12	0.013
σ_1	0.04	0.004	0.22	0.02

Source: Reproduced from Shoukri, M.M. et al., 2008. *BMC Medical Research Methodology,* 8(24), 1–11.

Note: The estimate of the common WSCV under the null is 0.034 (SE = 0.003). $\rho_{12} = 0.65$, $T_{wald} = -7.3$, LRT = 79, PM = −4.6 ($p < 0.001$ for all tests). 95% CI for ($\theta_1 - \theta_2$): (−0.12, − 0.07).

are provided in the accompanied CD. The measurements are the size of the brain ventricle relative to that of the patient's skull, and given by the ventricle-brain ratio (VBR) = (ventricle size/brain size) × 100. For a given scan, VBR was determined from measurements of the perimeter of the patient's ventricle together with the perimeter of the inner surface of the skull. These measurements were taken either: (1) from an automated pixel count (PIX) based on the images displayed on a television screen, or (2) a hand-held planimeter (PLAN) on a projection of the x-ray image. Table 4.7 summarizes the results. Clearly, all tests show that PIX has significantly lower WSCV than the PLAN ($p < 0.001$); that is better reproducibility.

Example 4.6

As a second example we illustrate the proposed methodologies by analyzing data from a biomedical study. The data are the on the gene expression measurement results of identical RNA preparations for two commercially available microarray platforms, namely, Affymerix (25-mer), and Amersham (30-mer). The RNA was collected from pancreatic PANC-1 cells grown in a serum-rich medium ("control") and 24 h following the removal of the serum ("treatment"). Three biological replicates (B1, B2, and B3) and three technical replicates (T1, T2, and T3) for the first biological replicate (B1) were produced by each platform. Therefore, for each condition (control and treatment) five hybridizations are conducted. The data set consists of 2009 genes that are identified as common across the platforms after comparing their Gene Bank IDs, and is normalized according to the manufacturer's standard software and normalization procedures. More details concerning this data set can be found in the original article by Tan et al. (2003).

The results presented are not restricted to the group of differentially expressed genes, and we used the "control" part of the data for both technical and biological replicates. The normalized intensity values are averaged for genes with multiple probes for a given Gene ID. Hence, we have a sample size of $n = 2009$ genes measured three times ($m = 3$) by each of the two platforms (or instruments). We have used the within-gene CV as a measure of reproducibility of a specific platform.

TABLE 4.8

Microarray Gene Expression Data Results ($n = 2009$ genes, $m = 3$ replicates)

(a) Technical Replicate

	Affymetrix ($l = 1$)		Amersham ($l = 2$)	
	Estimate	SE	Estimate	SE
μ_l	2759	150.5	3.74	0.22
ρ_l	0.94	0.002	0.99	0.0003
θ_l	0.58	0.03	0.25	0.015
σ_l	1603	17.88	0.93	0.01

Note: The estimate of the common WSCV under the null is 0.31 (SE = 0.014).
$\rho_{12} = 0.51$, $T_{wald} = 11.85$, LRT = 122, PM = −7.89 (p < 0.001 for all tests).
95% CI for ($\theta_1 - \theta_2$): (0.28, 0.39).

(b) Biological Replicate

	Affymetrix ($l = 1$)		Amersham ($l = 2$)	
	Estimate	SE	Estimate	SE
μ_l	2819	142.6	3.43	0.18
ρ_l	0.91	0.003	0.93	0.0025
θ_l	0.71	0.037	0.63	0.034
σ_l	2003.7	22.35	2.16	0.02

Note: The estimate of the common WSCV under the null is 0.67 (SE = 0.025).
$\rho_{12} = 0.50$, $T_{wald} = 2.35$, LRT= 8.56, PM = −9.04 (p < 0.02 for all tests).
95% CI for ($\theta_1 - \theta_2$): (0.014, 0.15).

The results of the data analyses are summarized in Table 4.7. Parameter estimates for both platforms, the estimated WSCV under the null hypotheses, as well as confidence interval of the difference of the two WSCVs are given in the table. We note that the correlation estimates remain the same under both hypotheses. Moreover, we note that the ICC (ρ) are quite high. Using benchmarks provided in Landis and Koch (1977a), both platforms produce substantially reproducible gene expression levels. Clearly, this is due to the large heterogeneity among the genes in the data set. Application of the LRT, WT, and the PM tests for testing the equality of two dependent WSCV show that the Amersham has significantly lower WSCV ($p < 0.001$), that is, better reproducibility for both the technical and biological replicates (Table 4.8).

4.11 Some Simulation Results

The theoretical properties of the test procedures discussed thus far are largely intractable in finite samples. Shoukri et al. (2008) took a Monte Carlo study to determine the levels of significance and powers of the proposed tests over a wide range of parameter values. Simulations were performed using programs written in MATLAB® (The MathWorks, Inc., Natic, MA).

The parameters of the simulation included the total number of subjects (n), the number of replications ($m_1 = m_2 = m$), and various values of (θ_1, θ_2 ρ_1, ρ_2, ρ_{12}). For each of 2000 independent runs of an algorithm constructed to generate observations from multivariate normal distribution, we estimated the true level of significance and power of the LRT, WT, Score and PM tests using a nominal level of significance 5% (two sided) for various combinations of parameters.

Tables 4.9 and 4.10 report the empirical significance levels based on 2000 simulated data sets for four (WT, Score, LRT, and PM) procedures for sample size of $n = 50$ and $n = 100$, respectively. It is seen that all procedures provide satisfactory significance levels at all parameter values examined. The empirical significance levels for smaller sample sizes ($n = 10$, 20, and 30) were also estimated. All test procedures provided empirical levels that are very close to the 5% nominal level (data not shown).

Tables 4.11 and 4.12 display empirical powers based on 2000 simulated data sets for WT and LRT in sample sizes $n = 30$ and 50, respectively. As alluded to earlier, the score test is excluded from the power Tables 4.11 and 4.12, because its simulated empirical power values were unacceptably low (as we show in Table 4.13). We observe that for all parameter values that WT and LRT provide almost identical values of power (Tables 4.11 and 4.12). Thus, although the LRT shows greater power at some parameter combinations than the WT, the difference is usually less than three percentage

TABLE 4.9

Empirical Significance Levels Based on 2000 Runs at Nominal Level 5% (Two Sided) for Testing $\theta_1 = \theta_2 = 0.15$ using the LRT, WT, Score, and PM for $n = 50$ Subjects and m Replicates

$n = 50$	$\rho = 0.4$			$\rho = 0.6$			$\rho = 0.7$		
ρ_{12}	0.1	0.2	0.3	0.1	0.3	0.5	0.1	0.4	0.6
$m = 2$									
WT	0.049	0.048	0.050	0.051	0.050	0.049	0.046	0.052	0.048
Score	0.057	0.051	0.053	0.055	0.055	0.058	0.051	0.058	0.054
PM	0.052	0.050	0.047	0.050	0.049	0.048	0.053	0.052	0.051
LRT	0.051	0.051	0.052	0.052	0.051	0.049	0.050	0.048	0.050
$m = 3$									
WT	0.048	0.046	0.049	0.052	0.049	0.050	0.048	0.047	0.049
Score	0.056	0.053	0.055	0.058	0.054	0.051	0.054	0.047	0.051
PM	0.053	0.052	0.050	0.049	0.049	0.052	0.052	0.050	0.052
LRT	0.050	0.047	0.051	0.053	0.047	0.051	0.049	0.045	0.048
$m = 5$									
WT	0.048	0.049	0.052	0.045	0.049	0.050	0.050	0.049	0.046
Score	0.054	0.050	0.051	0.051	0.053	0.054	0.049	0.048	0.056
PM	0.050	0.052	0.050	0.048	0.051	0.049	0.053	0.052	0.047
LRT	0.051	0.051	0.050	0.048	0.050	0.049	0.049	0.050	0.044

Note: We set $\rho_1 = \rho_2 = \rho$.

TABLE 4.10

Empirical Significance Levels Based on 2000 Runs at Nominal Level 5% (Two Sided) for Testing $\theta_1 = \theta_2 = 0.15$ using the LRT, WT, Score, and PM for $n = 100$ Subjects and m Replicates

n = 100	$\rho = 0.4$			$\rho = 0.6$			$\rho = 0.7$		
ρ_{12}	0.1	0.2	0.3	0.1	0.3	0.5	0.1	0.4	0.6
$m = 2$									
WT	0.049	0.048	0.051	0.049	0.050	0.045	0.046	0.050	0.051
Score	0.050	0.056	0.056	0.053	0.055	0.049	0.051	0.057	0.056
PM	0.049	0.048	0.052	0.049	0.049	0.050	0.050	0.051	0.051
LRT	0.048	0.044	0.051	0.044	0.050	0.042	0.042	0.048	0.050
$m = 3$									
WT	0.051	0.050	0.048	0.048	0.049	0.049	0.051	0.047	0.044
Score	0.051	0.049	0.048	0.050	0.054	0.053	0.057	0.052	0.056
PM	0.052	0.050	0.052	0.048	0.049	0.049	0.049	0.050	0.048
LRT	0.050	0.051	0.050	0.047	0.046	0.048	0.048	0.050	0.043
$m = 5$									
WT	0.050	0.049	0.052	0.049	0.048	0.050	0.051	0.050	0.046
Score	0.053	0.052	0.054	0.054	0.053	0.051	0.052	0.053	0.052
PM	0.050	0.049	0.053	0.049	0.051	0.050	0.049	0.050	0.047
LRT	0.049	0.050	0.052	0.048	0.050	0.049	0.047	0.051	0.045

Note: We set $\rho_1 = \rho_2 = \rho$.

TABLE 4.11

Empirical Power Based on 2000 Runs for Testing $\theta_1 = \theta_2$ using the LRT and WT for $n = 30$ Subjects

n = 30	$(\rho_1, \rho_2) = (0.7, 0.5)$ $(\theta_1, \theta_2) = (0.1, 0.2)$			$(\rho_1, \rho_2) = (0.6, .5)$ $(\theta_1, \theta_2) = (0.15, 0.2)$			$(\rho_1, \rho_2) = (0.5, 0.4)$ $(\theta_1, \theta_2) = (0.2, 0.3)$		
ρ_{12}	0.2	0.3	0.4	0.2	0.3	0.4	0.1	0.2	0.3
$m = 2$									
WT	0.92	0.93	0.94	0.30	0.33	0.32	0.55	0.50	0.51
LRT	0.94	0.95	0.96	0.35	0.33	0.34	0.60	0.56	0.54
$m = 3$									
WT	0.99	1.00	1.00	0.55	0.56	0.54	0.79	0.80	0.79
LRT	1.00	1.00	1.00	0.57	0.56	0.55	0.82	0.83	0.83
$m = 5$									
WT	1.00	1.00	1.00	0.80	0.82	0.84	0.96	0.95	0.97
LRT	1.00	1.00	1.00	0.81	0.82	0.85	0.97	0.97	0.98

TABLE 4.12

Empirical Power Based on 2000 Runs for Testing $\theta_1 = \theta_2$ using the LRT and WT for $n = 50$ Subjects

$n = 50$	$(\rho_1, \rho_2) = (0.7, 0.5)$ $(\theta_1, \theta_2) = (0.1, 0.2)$			$(\rho_1, \rho_2) = (0.6, .5)$ $(\theta_1, \theta_2) = (0.15, 0.2)$			$(\rho_1, \rho_2) = (0.5, 0.4)$ $(\theta_1, \theta_2) = (0.2, 0.3)$		
ρ_{12}	0.2	0.3	0.4	0.2	0.3	0.4	0.1	0.2	0.3
$m = 2$									
WT	0.99	0.98	0.99	0.47	0.50	0.49	0.75	0.72	0.74
LRT	0.99	0.99	0.99	0.49	0.52	0.51	0.77	0.77	0.78
$m = 3$									
WT	1.00	1.00	1.00	0.76	0.77	0.78	0.94	0.95	0.95
LRT	1.00	1.00	1.00	0.79	0.78	0.79	0.95	0.95	0.96
$m = 5$									
WT	1.00	1.00	1.00	0.94	0.93	0.95	1.00	0.99	1.00
LRT	1.00	1.00	1.00	0.95	0.95	0.96	1.00	0.99	1.00

points. We also conducted simulations to estimate the powers of the test statistics for smaller sample sizes ($n = 10$ and 20) (data not shown). We found that for some parameter combinations WT and LRT provided acceptable power especially if the distance between θ_1 and θ_2 is large, and showed greater power than both the Score and PM tests. The power of Score test was generally very low.

For selected parameter values, power levels of PM, WT, and the score tests for $n = 50$ subjects are given in Table 4.12. As already mentioned, the power of the score test is generally low. We note that the power of the WT is quite sensitive to the distance between θ_1 and θ_2. We note that the equality of the means and variances implies the equality of the WSCV, but the reverse is not true. This strong assumption might explain the relatively poor performance of the PM test, particularly when the means are not well separated.

TABLE 4.13

Empirical Power of PM, Score, and WT Based on 2000 Data Sets, $n = 50$ Subjects, $m = 3$ Replicates

(μ_1, μ_2)	(θ_1, θ_2)	ρ_1	ρ_2	ρ_{12}	PM	Score	WT
(10, 10)	(0.2, 0.3)	0.5	0.4	0.3	0.53	0.37	0.94
	(0.2, 0.4)	0.5	0.3	0.2	0.84	0.51	0.99
(8, 10)	(0.2, 0.3)	0.5	0.4	0.3	0.71	0.40	0.95
	(0.2, 0.4)	0.5	0.3	0.2	0.69	0.51	1.00
(6, 10)	(0.2, 0.3)	0.5	0.4	0.3	0.84	0.35	0.94
	(0.2, 0.4)	0.5	0.3	0.2	0.99	0.54	1.0
(5, 10)	(0.2, 0.3)	0.5	0.4	0.3	0.91	0.40	0.95
	(0.2, 0.4)	0.5	0.3	0.2	0.997	0.54	1.00

To assess the effect of nonnormality on the properties of the proposed test statistics, we generated data from a log-normal distribution, and evaluated the performance of the four procedures for 2000 simulated data sets. The empirical levels of the regression based PM test were quite close to the 5% nominal level, but the power was poor. However, the likelihood-based procedures (WT, LRT, and Score) did not preserve their nominal levels for the majority of the parameters combinations (data not shown).

EXERCISES

E4.1 The data of Example 4.4 are stored in a file in the CD (CT_SCAN_CHAPTER4). Find the Pearson's correlation between the two readings of each method. Which method has higher correlation?

E4.2 To help you with the data analysis in SAS, the scores of the first five subjects are given below:

```
input    PIX1    PIX3    PLAN1    PLAN3;
method = 1;  repl = 1;  y = PIX1;  output;
method = 1;  repl = 2;  y = PIX3;  output;
method = 2;  repl = 1;  y = PLAN1;  output;
method = 2;  repl = 2;  y = PLAN3;  output;
cards;
1.79   1.77    2.05       2.13
0.00   0.00    1.72       1.28
1.53   1.55    1.93       1.79
1.57   1.57    2.16       1.96
1.65   1.7     2.27       1.95
```

Find the ICC for each method. Find the within subject coefficients of variations for each method, using the moments estimators.

E4.3 Use the regression method to test the null hypothesis that the CVs are equal.

E4.4 Use the WT to test the above null hypothesis. Use the following strategies to find a reasonable estimate for ρ_{12}. The first step is to find

$$\hat{r}_1^2 = \frac{\text{cov}(PIX1, PLAN1)\text{cov}(PIX3, PLAN3)}{\text{cov}(PIX1, PIX3)\text{cov}(PLAN1, PLAN3)},$$

and

$$\hat{r}_2^2 = \frac{\text{cov}(PIX1, PLAN3)\text{cov}(PIX3, PLAN1)}{\text{cov}(PIX1, PIX3)\text{cov}(PLAN1, PLAN3)}.$$

The second step is to estimate ρ_{12} once by

$$\hat{\rho}_{12} = \frac{\sqrt{\hat{r}_1^2 + \hat{r}_2^2}}{2},$$

and second time, as suggested by Spearman (1910) by

$$\hat{\rho}_{12} = \sqrt{\hat{r}_1 \hat{r}_2}.$$

E4.5 Use the Delta method to find the approximate standard error of both the above-proposed estimates of ρ_{12}.

Appendix

Elements of Fisher's Information Matrix ($m_1 = m_2 = m$)

$$i_{33} = E\left(\frac{\partial^2 l}{\partial \sigma_1^2 \partial \sigma_1^2}\right) = \frac{n}{4\sigma_1^4}\left[2m + \frac{m^2}{w}\rho_{12}^2\right],$$

$$i_{34} = E\left(\frac{\partial^2 l}{\partial \sigma_1^2 \partial \sigma_2^2}\right) = -\frac{nm^2}{4\sigma_1^2\sigma_2^2 w}\rho_{12}^2,$$

$$i_{35} = E\left(\frac{\partial^2 l}{\partial \sigma_1^2 \partial \rho_1}\right) = -\frac{n(m-1)}{2\sigma_1^2}\left[\frac{1}{1-\rho_1} - \frac{1+(m-1)\rho_2}{w}\right],$$

$$i_{36} = i_{45} = 0,$$

$$i_{37} = E\left(\frac{\partial^2 l}{\partial \sigma_1^2 \partial \rho_{12}}\right) = -\frac{nm^2}{2w\sigma_1^2}\rho_{12},$$

$$i_{44} = E\left(\frac{\partial^2 l}{\partial \sigma_2^2 \partial \sigma_2^2}\right) = \frac{n}{4\sigma_2^4}\left[2m + \frac{m^2}{w}\rho_{12}^2\right],$$

$$i_{46} = E\left(\frac{\partial^2 l}{\partial \sigma_2^2 \partial \rho_2}\right) = -\frac{n(m-1)}{2\sigma_2^2}\left[\frac{1}{1-\rho_2} - \frac{1+(m-1)\rho_1}{w}\right],$$

$$i_{47} = E\left(\frac{\partial^2 l}{\partial \sigma_2^2 \partial \rho_{12}}\right) = -\frac{nm^2}{w}\frac{\rho_{12}}{2\sigma_2^2},$$

$$i_{55} = E\left(\frac{\partial^2 l}{\partial \rho_1^2}\right) = \frac{n(m-1)}{2}\left[\frac{1}{(1-\rho_1)^2} + (m-1)\frac{(1+(m-1)\rho_2)^2}{w^2}\right],$$

$$i_{56} = E\left(\frac{\partial^2 l}{\partial\rho_1\partial\rho_2}\right) = n(m-1)^2 m^2 \frac{\rho_{12}^2}{2w^2},$$

$$i_{57} = E\left(\frac{\partial^2 l}{\partial\rho_1\partial\rho_{12}}\right) = -\frac{n(m-1)m^2\rho_{12}}{w}\left[1+(m-1)\rho_2\right],$$

$$i_{66} = E\left(\frac{\partial^2 l}{\partial\rho_2^2}\right) = \frac{n(m-1)}{2}\left[\frac{1}{(1-\rho_2)^2} + (m-1)\frac{\left(1+(m-1)\rho_1\right)^2}{w^2}\right],$$

$$i_{67} = E\left(\frac{\partial^2 l}{\partial\rho_2\partial\rho_{12}}\right) = -n(m-1)\frac{m^2\rho_{12}}{w}\left[1+(m-1)\rho_1\right],$$

$$i_{77} = E\left(\frac{\partial^2 l}{\partial\rho_{12}^2}\right) = \frac{nm^2}{w}\left[1+2\rho_{12}^2\frac{m^2}{w^2}\right].$$

The matrix I_{22} is therefore given by

$$I_{22} = \begin{bmatrix} i_{33} & i_{34} & i_{35} & 0 & i_{37} \\ & i_{44} & 0 & i_{46} & i_{47} \\ & & i_{55} & i_{56} & i_{57} \\ & & & i_{66} & i_{67} \\ & & & & i_{77} \end{bmatrix}.$$

5

Measures of Agreement for Dichotomous Outcomes

5.1 Introduction

In Chapters 2 through 4, we examined indices by which one can determine the extent of agreement between two or more methods of evaluating test subjects, where none of the methods by itself can be accepted as a gold standard. The methods being compared may take a variety of forms such as: diagnostic devices, clinicians or examiners using the same procedure. In a test–retest situation, several independent applications of a given instrument or procedure by one rater are of interest as well. Although the raters may differ, test–retest studies of reliability generally have the same basic design. That is, methods of evaluation are used on one group of subjects at the same time or within an acceptable time interval. The assumption is made that the subjects being evaluated do not vary from one method to the other. The methods and models discussed therefore are applicable to evaluating continuous scale measurements (e.g., blood pressures, glucose level, bacterial counts, etc.). This and the following chapter are devoted to the methods of evaluating agreement among several raters for categorical assignments. Here are some examples.

In a methodological study conducted for the U.S. National Health Survey, two different questionnaire forms were administered to the same respondents within a 7–10 day interval in an effort to determine the degree of agreement between forms in the elicited response concerning the presence or absence of certain disease condition among respondents. In another study by Westlund and Kurland (1953), two neurologists reviewed the same set of selected medical records of potential MS patients and classified each of the individuals involved into one of four categories ranging from one certain to doubtful MS. The purpose here was to determine the extent to which trained neurologists agreed in their diagnosis of MS based on a medical record review.

The results of a test–retest study are usually summarized in a $c \times c$ table, where c is the number of categories into which a subject may be classified. This chapter is devoted to the case when $c = 2$.

TABLE 5.1

Basic 2×2 Table

		(1) Disease	(0) No Disease	Total
Rater 2 (X_2)	(1) Disease	n_{11}	n_{10}	$n_1.$
	(0) No disease	n_{01}	n_{00}	$n_2.$
	Total	$n._1$	$n._2$	n

One of the simplest situations is when we have a dichotomous classification (e.g., disease–no disease, absent–present, exposed–not exposed, etc.) resulting in a 2×2 table (Table 5.1).

A direct way of measuring what we call crude agreement is to compute the quantity $\text{var}(\hat{P}_0) = (n_{11} + n_{00})/n$, the proportion of the total that is in agreement. This index is called "simple-matching" coefficient, its estimated variance is given by $(\hat{P}_0) = P_0(1 - P_0)/n, 0 \leq P_0 \leq 1$.

This index has been heavily criticized, and in this chapter we reexamine this index, and review the attempts made by several researchers to improve it.

As we pointed out in the previous chapters, there are fundamental differences between agreement, reliability, and association. However, for the 2×2 table, agreement and association become indistinguishable under certain conditions. We further elaborate on this issue in the sections that follow.

5.2 Indices of Adjusted Agreement

The first index of adjusted agreement is the Jacquard coefficient which is the proportion of $(1, 1)$ matches in a set of comparisons that ignores $(0, 0)$ matches. This coefficient is estimated by

$$\hat{J} = \frac{n_{11}}{n_{11} + n_{10} + n_{01}}.$$

The estimated variance of Jacquard coefficient is given by

$$\hat{\text{Var}}(\hat{J}) = \frac{\hat{J}^2(1 - \hat{J})}{n_{11}} \quad 0 \leq \hat{J} \leq 1.$$

The second adjusted measure of similarity is the G coefficient, proposed by Holley and Guidford (1964) and Maxwell (1977) is given as

$$\hat{G} = \frac{[(n_{11} + n_{00}) - (n_{10} + n_{01})]}{n},$$

and its estimated variance is

$$\hat{V}ar(\hat{G}) = \frac{(1 - \hat{G}^2)}{n} \quad 1 \le \hat{G} \le -1.$$

The maximum value of $\hat{G} = 1$ indicates perfect similarity, and occurs when $n_{10} = n_{01} = 0$. The minimum value of $\hat{G}$ is -1 indicating perfect dissimilarity and occurs when $n_{11} = n_{22} = 0$. When $n_{11} + n_{00} = n_{10} + n_{01}$, then $\hat{G}$ is 0.0—a value that lies between the two extremes $(-1, 1)$.

A third adjusted measure of agreement is based on the concordance ratio,

$$C = \Pr\left[X_2 = 1 | X_1 = 1\right] = \frac{P\left(X_1 = 1, X_2 = 1\right)}{P\left(X_1 = 1\right)},$$

and the corresponding estimator is given by

$$\hat{C} = \frac{2n_{11}}{2n_{11} + n_{10} + n_{01}} = 1 - \frac{n_{10} + n_{01}}{2n_{11} + n_{10} + n_{01}}.$$

As can be seen, and similar to $\hat{j}$, the index $\hat{C}$ ignores the $(0, 0)$ cell, and gives twice the weight to the $(1, 1)$ cell. The maximum value of $\hat{C} = 1$ indicates perfect agreement, and occurs when $n_{10} = n_{01} = 0$. The minimum value of $\hat{C} = 0$ occurs when $n_{11} = 0$. The estimated variance of $\hat{C}$ is

$$\hat{V}ar(\hat{C}) = \frac{(1 - \hat{C})(2 - \hat{C})\hat{C}^2}{2n_{11}}.$$

$$= \frac{(1 - \hat{C})(2 - \hat{C})\hat{C}}{2n\hat{\pi}}.$$

Here $\hat{\pi} = (2n_{11} + n_{10} + n_{01})/2n$ is the maximum likelihood estimator of the probability $P(X_j = 1)$.

The intraclass correlation is perhaps the most widely known measure of similarity in a 2×2 table. It is estimated by

$$\hat{\rho} = \frac{4n_{11}n_{00} - \left(n_{10} + n_{01}\right)^2}{\left(2n_{11} + n_{10} + n_{01}\right)\left(2n_{00} + n_{10} + n_{01}\right)}$$

(see Hannah et al., 1983; Donner and Eliasziw, 1992).

Similar to the $\hat{G}$ index, the maximum value of $\hat{\rho}$ is 1 indicating perfect similarity, and occurs when $n_{10} = n_{01} = 0$. The minimum value of $\hat{\rho}$ is -1 that indicates perfect dissimilarity and occurs when $n_{11} = n_{00} = 0$. When $(n_{11}n_{00})^{1/2} = (n_{10} + n_{01})/2$, then $\hat{\rho}$ is 0.0. The estimated variance of $\hat{\rho}$ is:

$$\text{va}\hat{\text{r}}(\hat{\rho}) = \frac{1 - \hat{\rho}}{n}\left[(1 - \hat{\rho})(1 - 2\hat{\rho}) + \frac{\hat{\rho}(2 - \hat{\rho})}{2\hat{\pi}(1 - \hat{\pi})}\right]. \tag{5.1}$$

There are other measures of similarities, however, they are used to measure associations between the two sets of ratings, and as such they should not be used as measures of agreement. These are

1. *Odds ratio:* $\hat{\Psi} = (n_{11}n_{00}) / (n_{10}n_{01})$ and its estimated variance is

$$\text{va}\hat{\text{r}}(\hat{\Psi}) = \hat{\Psi}^2 A(n),$$

where

$$A(n) = n_{11}^{-1} + n_{10}^{-1} + n_{01}^{-1} + n_{00}^{-1}.$$

2. *Yule's coefficient:* $\hat{Y} = \hat{\Psi} - 1/\hat{\Psi} + 1$.

The estimated variance of Yule's coefficient is $\text{va}\hat{\text{r}}(\hat{Y}) = (1 - \hat{Y}^2)A(n)/4$, and a modified version of Yule's coefficient

$$\hat{m} = \frac{\left(1 + \hat{Y}\right)^{1/2} - \left(1 - \hat{Y}\right)^{1/2}}{\left(1 + \hat{Y}\right)^{1/2} + \left(1 - \hat{Y}\right)^{1/2}},$$

whose estimated variance is given by

$$\text{va}\hat{\text{r}}(\hat{m}) = (1 - \hat{m}^2)^2 A(n)/16.$$

Earlier, approaches to studying interrater agreement focused on the observed proportion of agreement or the "simple matching coefficient" denoted by P_o. This statistics does not allow for the fact that a certain amount of agreement can be expected on the basis of chance alone and may occur even if there were no systematic tendency for the raters to classify the subjects similarly. In the next section, we investigate chance corrected measure of agreement in the 2×2 classification.

5.3 Cohen's Kappa: Chance Corrected Measure of Agreement

Cohen (1960) proposed kappa as a chance-corrected measure of agreement, to discount the observed proportion of agreement by the expected level of agreement, given the observed marginal distributions of the rater's responses under the assumption that the rater's reports are statistically independent. Cohen assumed that there are two raters, who rate n subjects into one of two mutually exclusive and exhaustive nominal categories. These raters classify subjects independently. Since the observed agreement

$$P_o = \frac{n_{11} + n_{00}}{n},$$

and

$$P_e = \left(\frac{n_{1\cdot}}{n}\right)\left(\frac{n_{\cdot 1}}{n}\right) + \left(\frac{n_{2\cdot}}{n}\right)\left(\frac{n_{\cdot 2}}{n}\right)$$

is the proportion agreement expected by chance, the Kappa coefficient proposed by Cohen is

$$\hat{\kappa} = \frac{P_0 - P_e}{1 - P_e}. \tag{5.2}$$

Fleiss et al. (1969) provided an approximate asymptotic expression for the estimated variance of $\hat{\kappa}$, given as

$$\text{var}(\kappa) = \frac{1}{n(1 - P_e)^2}\left(\sum_{I=1}^{2} \hat{P}_{ii}\left[1 - \left(\hat{P}_{i\cdot} + \hat{P}_{\cdot i}\right)\left(1 - \hat{\kappa}\right)\right]^2 \right.$$

$$\left. + \left(1 - \hat{\kappa}\right)^2 \sum_{i \neq j}^{2} \hat{P}_{ii}\left(\hat{P}_{i\cdot} + \hat{P}_{\cdot j}\right)^2 - \left\{\hat{\kappa} - P_e\left(1 - \hat{\kappa}\right)\right\}^2 \right). \tag{5.3}$$

5.4 Intraclass Kappa

Bloch and Kraemer (1989) introduced the ICC as an alternative version of Cohen's kappa, under the assumption that each rater is characterized by the same marginal probability of positive diagnosis. The intraclass version of the kappa statistic is algebraically equivalent to Scott's (1955) index of agreement. The intraclass kappa was defined by Bloch and Kraemer (1989) for situations consisting of blinded binary diagnoses on each of the n subjects by two raters. It is assumed that the two ratings on each subject are interchangeable; that is, in the population of subjects, the two ratings for each subject have a distribution that is invariant under the permutation of the raters (i.e., absence of interrater bias). We derive the intraclass kappa as follows.

Let X_{ij} denote the rating for the ith subject by jth rater, $i = 1, ..., n, j = 1, 2,$ and conditional on the ith subject let $P_r(X_{ij} = 1 | p_i) = p_i$ be the probability of positive rating (diagnosis). Over the population of subjects, let $E(p_i) = \pi$ and $\text{var}(p_i) = \rho\pi(1 - \pi)$. Therefore, unconditionally, $E(X_{ij}) = \pi$, $\text{var}(X_{ij}) = \pi(1 - \pi)$, and $\text{cov}(X_{i1}, X_{i2}) = \rho\pi(1 - \pi)$, and the intraclass correlation is defined as

$$\rho = \frac{\text{cov}(X_{i1}, X_{i2})}{\sqrt{\text{var}(X_{i1})\text{var}(X_{i2})}}.$$

The underlying probabilistic model (known as the common correlation model) of the above experiment is summarized in Table 5.2.

Since

$$P_0 = P_1(\rho) + P_3(\rho) = \pi^2 + (1 - \pi)^2 + 2\rho\pi(1 - \pi),$$

and

$$P_e = \pi^2 + (1 - \pi)^2,$$

then

$$\kappa_1 = \frac{P_0 - P_e}{1 - P_e} = \frac{2\rho\pi(1 - \pi)}{1 - \pi^2 - (1 - \pi)^2} = \rho.$$

Thus, we have established equivalence of intraclass kappa to the ICC.

TABLE 5.2

Distribution of Responses under the Diagnostic Categories

x_{i1}	x_{i2}	Obs. Freq.	$P(X_{i1} = x_1, X_{i2} = x_2)$
1	1	n_{11}	$\pi^2 + \rho\pi(1 - \pi) \equiv P_1(\rho)$
1	0	n_{10}	$\pi(1 - \pi)(1 - \rho) \equiv P_2(\rho)$
0	1	n_{01}	$\pi(1 - \pi)(1 - \rho) \equiv P_2(\rho)$
0	0	n_{00}	$(1 - \pi)^2 + \rho\pi(1 - \pi) \equiv P_3(\rho)$

The MLE $\hat{\pi}$ and $\hat{\kappa}$ for π and κ are, respectively

$$\hat{\pi} = \frac{(2n_{11} + n_{10} + n_{01})}{(2n)}$$

and

$$\hat{\kappa}_1 = \frac{4(n_{11}n_{00} - n_{10}n_{01}) - (n_{10} - n_{01})^2}{(2n_{11} + n_{10} + n_{01})(2n_{00} + n_{10} + n_{01})}. \tag{5.4}$$

This is identical to $\hat{\kappa}_1$ and its asymptotic variance is given by:

$$\text{var}(\hat{\kappa}_1) = \frac{1 - \kappa_1}{\kappa_1}\left[(1 - \kappa_1)(2 - \kappa_1) + \frac{\kappa_1(2 - \kappa_1)}{2\pi(1 - \pi)}\right] \tag{5.5}$$

If the formula for the ICC for continuous data under the one-way random effects model is applied to the 0–1 data, then the estimate $\hat{\kappa}_1$ is obtained. Under certain conditions, we assume that is asymptotically normally distributed with mean κ_1 and standard error $SE(\hat{\kappa}_1) = \sqrt{\text{var}(\hat{\kappa}_1)}$, and the $100(1 - \alpha)\%$ confidence interval is given by $\hat{\kappa}_1 \pm Z_{1-\alpha/2}SE(\hat{\kappa}_1)$, where $Z_{1-\alpha/2}$ is the $100(1 - \alpha/2)$ percentile point of the standard normal distribution. This confidence interval has acceptable properties in large samples (Bloch and Kraemer, 1989; Donner and Eliasziw, 1992).

Donner and Eliasziw (1992) proposed confidence interval based on a chi-square goodness-of-fit statistics that is appropriate in small samples. Their approach is based on equating the computed one-degree-of-freedom chi-square statistics to an appropriately selected critical value, and solving for the two roots of kappa. Specifically, to test $H_0 : \kappa_1 = \kappa_0$, one refers

$$X_G^2 = \frac{\left[n_{11} - n\hat{P}_1(\kappa_0)\right]^2}{n\hat{P}_1(\kappa_0)} + \frac{\left[n_{10} + n_{01} - 2n\hat{P}_2(\kappa_0)\right]^2}{2n\hat{P}_2(\kappa_0)} + \frac{\left[n_{00} - n\hat{P}_3(\kappa_0)\right]^2}{n\hat{P}_3(\kappa_0)} \tag{5.6}$$

to the chi-square distribution with one degree of freedom at the chosen level of significance α. The $\hat{P}_l(\kappa_0)$ are obtained by replacing π by $\hat{\pi}$ in $P_1(\kappa_0)$, $l = 1,2,3$. Using this approach the upper and lower limits of a 95% confidence interval for κ_1 are given respectively as

$$\kappa_u = \left(\frac{1}{9}y_3^2 - \frac{1}{3}y_2\right)^2\left(\cos\frac{\theta + 5\pi}{3} + \sqrt{3}\sin\frac{\theta + 2\pi}{3}\right) - \frac{1}{3}y_3,$$

$$\kappa_L = 2\left(\frac{1}{9}y_3^2 - \frac{1}{3}y_2\right)^{\frac{1}{2}}\cos\frac{\theta + 5\pi}{3} - \frac{1}{3}y_3.$$

Note that $\pi = 22/7$ in the upper and lower limits. Moreover,

$$\theta = \cos^{-1}\frac{V}{W}, V = \frac{1}{27}y_3^2 - \frac{1}{6}(y_2 y_3 - 3y_1),$$

$$W = \left(\frac{1}{9}y_3^2 - \frac{1}{3}y_2\right)^{\frac{3}{2}},$$

and

$$y_1 = \frac{\left\{n_{10} + n_{01} - 2n\hat{\pi}(1-\hat{\pi})\right\}^2 + 4n^2\hat{\pi}^2(1-\hat{\pi})^2}{4n\hat{\pi}^2(1-\hat{\pi})^2(n+3.84)} - 1,$$

$$y_2 = \frac{(n_{10} + n_{01})^2 - 4(3.84)n\hat{\pi}(1-\hat{\pi})\left\{1 - 4\hat{\pi}(1-\hat{\pi})\right\}}{2n\hat{\pi}^2(1-\hat{\pi})^2(n+3.84)} - 1,$$

$$y_3 = \frac{n_{10} + n_{01} + 3.84\left\{1 - 2\hat{\pi}(1-\hat{\pi})\right\}}{\hat{\pi}(1-\hat{\pi})(n+3.84)} - 1.$$

5.5 2 × 2 Kappa in Context of Association

For the reliability kappa, the two ratings per subject are meant to be inter-changeable. In this context, ratings per subjects do not depend on the order or rating. However, if the ratings are not interchangeable, then the context is association, not agreement (Bloch and Kraemer, 1989).

When there are two independent, possibly different, ratings per subject, X_1 (with responses 1 and 0) and X_2 (with responses 1 and 0), for subject i let, $p_1 = P_r(X_1 = 1)$ and $p_2 = P_r(X_2 = 1)$. In this case, the theoretical model for 2×2 data is summarized in Table 5.3.

Since

$$\text{cov}(X_1, X_2) = E(X_1 X_2) - E(X_1)E(X_2),$$

then

$$\rho\sqrt{p_1 p_2 q_1 q_2} = E(X_1 X_2) - p_1 p_2.$$

TABLE 5.3

Distribution of Responses under Association Kappa

		Rater 1 Responses		Total
		$x_1 = 1$	$x_1 = 0$	
Rater 2 responses	$x_2 = 1$	$p_1 p_2 + \tau$	$p_2 q_1 - \tau$	p_2
	$x_2 = 0$	$p_1 q_2 - \tau$	$q_1 q_2 + \tau$	$1 - p_2 = q_2$
	Total	p_1	$1 - p_1 = q_1$	1

Therefore,

$$E(X_1 X_2) = p_1 p_2 + \rho \sqrt{p_1 p_2 q_1 q_2},$$

implying that

$$\tau = \rho (p_1 p_2 q_1 q_2)^{1/2}.$$

Here ρ is the correlation coefficient between X_1 and X_2. The chance corrected agreement in the context of association is therefore given as

$$\kappa_a = \frac{2\rho (p_1 p_2 q_1 q_2)^{1/2}}{p_1 q_2 + p_2 q_1}. \tag{5.7}$$

If the two raters are unbiased relative to each other (i.e., $p_1 = p_2$), then

$$\kappa_a = \rho.$$

The MLE of the model parameters are

$$\hat{p}_1 = \frac{(n_{11} + n_{01})}{n} \quad \hat{p}_2 = \frac{(n_{11} + n_{10})}{n},$$

and

$$\hat{\kappa}_a = \frac{2(n_{11}n_{00} - n_{10}n_{01})}{(n_{11} + n_{10})(n_{00} + n_{10}) + (n_{11} + n_{01})(n_{00} + n_{01})}.$$

For convenience, Table 5.3 will be written as in Table 5.3a.

The large sample variance of $\hat{\kappa}_a$ was given by Bloch and Kraemer (1989) and Shoukri et al. (1995) as

$$\text{Var}(\hat{\kappa}_a) = \frac{4 p_1 p_2 q_1 q_2}{(p_1 q_2 + p_2 q_1)^2} \phi(\rho) \tag{5.8}$$

TABLE 5.3a

Cross Classification for Association Kappa

		x_1		
		1	0	Total
x_2	1	p_{11}	p_{10}	p_2
	0	p_{01}	p_{00}	q_2
	Total	p_1	q_1	1

where

$$\phi(\rho) = 1 + 4U_x U_y \rho - \left(1 + 3U_x^2 + 3U_y^2\right)\rho^2 + 2U_x U_y \rho^3,$$

$$\rho = \frac{(p_1 q_2 + p_2 q_1)\kappa_a}{2(p_1 p_2 q_1 q_2)^{1/2}},$$

$$U_x = \frac{\left(\dfrac{1}{2} - p_1\right)}{\sqrt{p_1 q_1}},$$

and

$$U_y = \frac{\left(\dfrac{1}{2} - P_2\right)}{\sqrt{p_2 q_2}}.$$

A consistent estimator of $\mathrm{Var}(\hat{\kappa}_a)$ is obtained on replacing the parameters by their MLE $(\hat{p}_1, \hat{p}_2, \hat{\kappa}_a)$.

A test of interrater bias (or $P_1 = P_2$) may be conducted using McNemar's test (1947). Since the proportion of subjects classified as 1s by the first rater is estimated by

$$\hat{p}_1 = \frac{n_{11} + n_{01}}{n}$$

and the proportion classified as 1s by the second rater is estimated by

$$\hat{p}_1 = \frac{n_{11} + n_{01}}{n}.$$

Then, the difference between the proportions is

$$\hat{D} = \hat{p}_1 - \hat{p}_2 = \frac{n_{01} - n_{10}}{n} = \frac{n_{1\cdot}}{n} - \frac{n_{\cdot 1}}{n},$$

does not depend on n_{11}.

Testing for marginal homogeneity is equivalent to testing the null hypothesis: $H_0 : P_1 = P_2$, or equivalently $H_0 : P_{10} = P_{01}$ or $H_0 : D = P_{10} - P_{01} = 0$.
Here

$$\text{var}(\hat{D}) = \text{var}(\hat{P}_1) + \text{var}(\hat{P}_2) - 2\text{cov}\left(\hat{P}_1, \hat{P}_2\right),$$

but

$$\text{var}(\hat{P}_j) = \frac{P_j(1 - P_j)}{n}(j = 1, 2),$$

and

$$\text{cov}(\hat{P}_1, \hat{P}_2) = \frac{1}{n}\left[P_{11}(1 - P_{11}) - P_{11}(P_{10} + P_{01}) - P_{10}P_{01}\right]$$

$$= \frac{1}{n}\left[P_{11}(P_{10} + P_{01} + P_{22} - P_{10} - P_{01}) - P_{10}P_{01}\right]$$

$$= \frac{1}{n}\left[P_{11}P_{22} - P_{10}P_{01}\right].$$

Hence, under the null hypothesis $P_{10} = P_{01}$, we obtain

$$n\,\text{var}(\hat{D}) = P_{10} + P_{01},$$

from which

$$V = \text{var}(\hat{D}) = \frac{P_{10} + P_{01}}{n}.$$

When the sample size is large, we have, under H_0, $\hat{D} \sim N(0, V)$.
V is estimated by

$$\hat{V} = \frac{(n_{10} + n_{01})}{n^2}.$$

Hence,

$$Z^2 = \hat{D}^2/\hat{V} = \frac{(n_{10} - n_{01})^2}{n_{10} + n_{01}}$$

has a chi-square distribution with one degree of freedom.

Edwards (1948) suggested that the statistic

$$X^2 = \frac{\left(\left|n_{01} - n_{10}\right| - 1\right)^2}{n_{01} + n_{10}} \tag{5.9}$$

be used to test $H_0 : P_1 = P_2$. The value of X^2 should be referred to tables of chi-square with one degree of freedom. If X^2 is large, the inference can be made that the two raters are biased relative to each other.

5.6 Stratified Kappa

There are circumstances when the marginal probability of classification of a particular subject may depend on one or more confounding variables. For example, a physician rating whether a patient has particular disease symptoms may be influenced by the overall severity of the disease. On the basis of these confounders, one may want to assess the interrater agreement with subjects grouped into strata. Barlow et al. (1991) discussed several approaches for evaluating stratified agreement, assuming that the underlying kappa is common across strata but that the probability structure for each table may differ. The problem then arises of combining the strata to yield a summary "stratified" kappa statistic. They consider several weighting schemes and compare them in a simulation study. Their suggested weighting schemes are (1) equal weighting; (2) weighting by the relative sample size of each table; or (3) weighting by the inverse variance.

1. Equal weighting, although computationally simple, will likely have poor properties. Barlow et al. (1991) denoted this measure by κ_{ave}. If there are M strata, $\kappa_{ave} = (1/M)\sum_m \hat{\kappa}_m$ and $\text{var}(\kappa_{ave}) = (1/M^2)\sum_m \text{var}(\hat{\kappa}_m)$, where $\text{var}(\hat{\kappa}_m)$ is given by Equation 5.5.
2. Weighting by relative sample size n_m gives

$$\hat{\kappa}_s = \left(\sum_{m=1}^{M} n_m\right)^{-1} \sum_{m=1}^{M} n_m \hat{\kappa}_m.$$

The variance of this estimator is given by

$$\text{var}(\hat{\kappa}_s) = \left(\sum_{m=1}^{M} n_m\right)^{-2} \sum_{m=1}^{M} n_m^2 \, \text{var}(\hat{\kappa}_m).$$

3. The third method weights by the inverse of the variance of $\hat{\kappa}_m$.

Defining

$$W_m = \frac{\text{var}^{-1}(\hat{\kappa}_m)}{\displaystyle\sum_{m=1}^{M}\text{var}^{-1}(\hat{\kappa}_m)}, \qquad (5.10)$$

the stratified estimate of kappa is $\hat{\kappa}_V = \sum_{m=1}^{M} W_m \hat{\kappa}_m$. Its variance is given by

$$\text{var}(\hat{\kappa}_V) = \left(\sum_{m=1}^{M}\text{var}^{-1}(\hat{\kappa}_m)\right)^{-1}. \qquad (5.11)$$

Barlow et al. (1991) showed that weighting by the inverse of the variance estimator of $\hat{\kappa}$ given by Equation 5.5 results in estimates with large bias and inappropriate test size, and that method (2) has a higher variance.

Example 5.1

Prostate cancer is the most prevalent cancer in men. The approach to treatment varies and is dependent on the extent of cancer at the time of diagnosis. Although new imaging techniques have been developed over the past 15–25 years to increase staging accuracy and thereby lead to better treatment decisions, the increasing need for cost containment has raised questions about the value of these approaches. Computed tomography was initially used to stage prostate cancer, but since it cannot identify intrinsic prostate disease, it has been replaced by endorectal, or transrectal, ultrasonography for diagnosis and localized staging, and, in many institutions, by MRI for staging. It is well-known that the costs of these techniques are high. This identification of an accurate, not costly diagnostic technique is important for quality of care and cost containment. A modified TNM (tumor, node, and metastasis) staging system was to be used to categorize MRI, ultrasound and pathological finding. The results are summarized in Tables 5.4 and 5.5.

TABLE 5.4

Ultrasonography versus Pathological Analysis for Prostate Cancer Differentiation

	Stage in Pathological Study		
Stage in Ultrasound	**Advanced**	**Localized**	**Total**
Advanced	45	50	95
Localized	60	90	150
Total	105	140	245

TABLE 5.5

MRI versus Pathological Analysis for Prostate Cancer Differentiation

	Stage in Pathological Study		
Stage in MRI	**Advanced**	**Localized**	**Total**
Advanced	51	28	79
Localized	30	88	118
Total	81	116	197

The SAS code to analyze the data is given below.

```
data tumor1;
input instrument $ ULS $ Pathology $ count;
cards;
ULS      adv         adv       45
ULS      adv         local     50
ULS      local       adv       60
ULS      local       local     90

proc freq data = tumor1;
weight count;
tables ULS*pathology/agree;
run;
```

This gives an estimate of kappa 0.0723 with standard error 0.0638

```
data tumor2;
input instrument $ MRI $ Pathology $ count;
cards;
MRI      adv         adv       51
MRI      adv         local     28
MRI      local       adv       30
MRI      local       local     88
;
proc freq data = tumor2;
weight count;
tables MRI*pathology/agree;
run;
```

The estimate of kappa is 0.3897 and the standard error is 0.0667. For illustration, we have the following estimates by the three methods:

1. Simple average $\hat{\kappa}_{ave}$ = 0.231(0.046).
2. Weighting by the relative sample size, $\hat{\kappa}_s$ = (245(0.0723) + 197(0.3897)/ 245 + 197) = 0.214, and its estimated standard error is 0.046.
3. Weighting by the inverse of the variance, var_2^{-1} = 245.67, var_2^{-1} = 224.77. Hence w_1 = (245.67/245.67 + 224.77), $w_2 = 1 - w_1$. From which, $\hat{\kappa}_v$ = 0.224 and its standard error is 0.046.

TABLE 5.6

Detection and Localization of Lesions by Ultrasound, MRI, or Both, According to Lesion Size on Pathological Examination (Hypothetical Data)

Lesion Size Mm	Missed by Both	Seen by Both	Seen by MRI Only	Seen by Ultrasound Only	Total
1–5	40	20	15	10	85
6–10	29	70	14	10	123
11–15	10	12	7	7	36
16–20	3	24	2	3	32
21–25	0	10	1	1	12
≥26	1	8	2	1	12
Total	83	144	41	32	300

It should be noted that the three methods produce almost identical pooled estimate and identical standard errors.

Example 5.2

It has been reported in many investigations, for example, by Rifkin et al. (1990) that the ability to identify lesions varied directly with size but minimally with the plane of imaging. They reported that ultrasonography has identified 53% of all lesions ≤1 cm in the antero-posterior dimension, whereas it identified 72% of the lesions that were larger than 1 cm. Hypothetical data showing the corresponding percentages for MRI were 56% for the smaller lesions and 71% for the larger ones (Table 5.6). We use these data to compute estimates stratified kappa, where the strata are the different sizes of lesions. The SAS code is given after Table 5.6.

```
data lesion;
input lesion $ MRI $ ULT_SOUND $ count;
cards;
1_5        yes        yes        20
1_5        yes        no         15
1_5        no         yes        10
1_5        no         no         40
6_10       yes        yes        70
6_10       yes        no         14
6_10       no         yes        10
6_10       no         no         29
11_15      yes        yes        12
11_15      yes        no         7
11_15      no         yes        7
11_15      no         no         10
16_20      yes        yes        24
16_20      yes        no         2
16_20      no         yes        3
16_20      no         no         3
21_25      yes        yes        10
```

```
21_25      yes      no       1
21_25      no       yes      1
21_25      no       no       0
+26        yes      yes      8
+26        yes      no       2
+26        no       yes      1
+26        no       no       1
;
proc sort;
by lesion;
proc freq;
by lesion;
weight count;
tables MRI*ULT_SOUND /agree;
run;
```

The results are summarized in Table 5.7.

5.7 Conceptual Issues

Despite the popularity of kappa as a measure of agreement between raters, the readers should be made aware of its limitations and disadvantages. For example, it has been argued that this index depends heavily on the true prevalence of the condition being diagnosed. In the evaluation of diagnostic markers, it is well-known that certain test that seems to have high sensitivity and specificity may have low predictive accuracy when prevalence of the disease is low. Analogously, two raters who seem to have high agreement may nevertheless produce low values of kappa. This was clarified by Kraemer (1979) who showed how the prevalence of the condition would alter the results of kappa despite constant values of accuracy for each rater. Thompson and Walter (1988) extended the argument made by Kraemer showing that if

TABLE 5.7

Summary Analyses of the Data of Example 5.2

Lesion	Kappa	ASE
1_5	0.3796	0.1020
6_10	0.5615	0.0791
11_15	0.2198	0.1628
16_20	0.4521	0.2064
21_25	−0.0909	0.0640
+26	0.2500	0.3172

the errors of the two binary classifications are assumed to be independent, the role of kappa can be extended as an index of validity. In this case kappa may be written as

$$\kappa_{tw} = \frac{2\theta(1-\theta)(1-\alpha_1-\beta_1)(1-\alpha_2-\beta_2)}{\pi_1(1-\pi_2)+\pi_2(1-\pi_1)}, \tag{5.12}$$

where
  π_i = Proportion classified as having the characteristic according to the ith rater ($i = 1, 2$);
  $1 - \alpha_i$ = Specificity for the ith rater;
  $1 - \beta_i$ = Sensitivity for the ith rater;
  θ = The true proportion having the characteristic (the true prevalence)
(Table 5.8).

Note that $\pi_i = \theta(1 - \beta_i) + (1 - \theta)\alpha_i$. When $\alpha_1 = \alpha_2$, and $\beta_1 = \beta_2$, then κ_{tw} reduces to the expression given by Kraemer (1979). Expression (Equation 5.12) demonstrates the heavy dependence of kappa on θ, even if $\alpha_i = \beta_i = 0$ ($i = 1, 2$).

Since neither θ, nor α_i, β_i are available in practice, the importance of Equation 5.12 stems not from any potential for application in actual studies but instead from the insight it provides into the dependence of kappa on the true prevalence of the condition. Shrout et al. (1981) argued that the dependence of kappa on the prevalence may be a desired property. However, as can be seen from Equation 5.12, this strong dependence of kappa on the true prevalence of the characteristic of interest complicates its interpretation as an index of quality of measurement. For, it would seem especially difficult to compare two or more kappa values when the true prevalence for the conditions compared may differ.

Other controversial issues were discussed by Feinstein and Cicchetti (1990 I) and Feinstein and Cicchetti (1990 II) who noted that in the absence of "gold standard" reference measurement is not available for calculating sensitivity and specificity, and true prevalence is not determined. For these reasons, raters' variability is better assessed by the kappa coefficient. During

TABLE 5.8

Cross Classification of Rater (i) by Gold Standard

	Gold		
Rater (i)	D^+	D^-	**Total**
T^+	$\theta(1-\beta_i)$	$\alpha_i(1-\theta)$	π_i
T^-	$\theta\beta_i$	$(1-\theta)(1-\alpha_i)$	$1-\pi_i$
	θ	$1-\theta$	

those assessments, the investigators sometimes find a striking paradox: despite a relatively high result for the crude proportion of interrater agreement, the corresponding value of kappa may be relatively low. If one inspects the expression of the chance corrected agreement,

$$\kappa = \frac{P_0 - P_e}{1 - P_e}.$$

It is clear that for fixed P_0, kappa gets its highest value when P_e is as small as possible. Consider the data in Table 5.9.

It can be seen from the table that with different values of P_e, the kappa for identical values of P_0 can be more than twice as high in one instance as compared to the other. Feinstein and Cicchetti (1990) provided the following explanations. A low value of kappa despite a high value of P_0 will occur only when the marginal totals are highly symmetrically unbalanced (HSU). The HSU situations occurs when n_1 is very different from n_2 or when $n_{.1}$ is very different from $n_{.2}$. Perfect balance occurs when $n_1 = n_2$ or when $n_{.1} = n_{.2}$. As an example, we consider the data in Table 5.10.

Here, $P_0 = 0.84$ is indicating high observed agreement. However,

$$P_e = ((43)(45)/2500) + ((7)(5)/2500) = 0.79, \text{ and } \kappa = (0.84 - 0.79/1 - 0.79) = 0.24,$$

which is well below the observed agreement.

As can be seen from Table 5.9, this paradox is caused by the marked difference between $n_1 = 43$, and $n_2 = 7$ or because of the marked difference between $n_{.1} = 45$, and $n_{.2} = 5$.

A second paradox occurs when unbalanced marginal totals produce higher values of kappa than more balanced total. This situation occurs when n_1 is much larger than n_2, while $n_{.1}$ is much smaller than $n_{.2}$, or vice versa. This situation, which produces "asymmetrical unbalanced marginals" is illustrated in Table 5.10a.

$(n_1 > n_2$, but $n_{.1} < n_{.2})$

Here, $P_0 = 0.84$, $P_e = 0.50$, and $\kappa = 0.68$ which is much higher than kappa is obtained from symmetrically unbalanced table.

The issue of dependence of kappa on the prevalence has also been discussed in the radiology literature. We refer the readers to the papers by Albaum et al. (1996) and Kundel and Polansky (2003). For example, the data in Table 5.10b shows the cross classifications by two readers of 150 images for the presence or absence of a specific disease.

As we have shown, the observed agreement $P_0 = (5 + 130/170) = 0.79$, and the agreement expected by chance is $P_e = ((20)(25) + (145)(150)/(170)^2) = 0.77$. Hence,

$$\kappa = \frac{0.79 - 0.77}{1 - 0.77} = 0.09.$$

TABLE 5.9

Dependence of Kappa on Chance Agreement

P_0	P_e	Kappa
0.85	0.50	0.70
0.85	0.78	0.32

TABLE 5.10

Demonstrating a Symmetric Unbalance

	Rater (1)		
Rater (2)	Yes	No	Total
Yes	40	5	$n_{1.} = 45$
No	3	2	$n_{2.} = 5$
Total	$43 = n_{.1}$	$7 = n_{.2}$	40

TABLE 5.10a

Demonstrating Asymmetrical Unbalance

	Rater (1)		
Rater (2)	Yes	No	Total
Yes	21	6	27
No	2	21	23
Total	23	27	50

TABLE 5.10b

Rating of 150 Images by Two Raters

	Second Reader		
First Reader	+	−	Total
+	5(a)	15(b)	20
−	20(c)	130(d)	150
Total	25	145	$170 = n$

Kundel and Polansky (2003) attributed the paradoxical difference between the good overall agreement ($p_0 = 0.79$) and the weak chance-corrected agreement ($\kappa = 0.09$), to the high prevalence of negative cases.

Cicchetti and Feinstein (1990) suggested that when the investigators report the results of the studies of agreement they should respect, positive agreement, and negative agreement as well. This approach provides more details about where disagreement occurs. The proportions of positive and negative agreements are, respectively:

$P^+ = 2a/(n + a - d) = 0.22$, $P^- = 2d/(n - a + d) = 0.54$. Moreover, the prevalence index

$$P_{index} = \frac{|a - d|}{n} = \frac{|5 - 130|}{170} = 0.74$$

and the bias index is

$$b_{index} = \frac{|b - c|}{n} = \frac{5}{170} = 0.03.$$

The values 0.54 and 0.74, for the proportions of negative agreement and the prevalence index respectively, help understand the data. If the kappa alone was reported we may conclude that there is a weak agreement between raters. On the other hand, we realize that the low value of kappa was the result of a high prevalence of the negative cases.

Byrt et al. (1993) proposed using a statistic called "Prevalence Adjusted Bias Adjusted Kappa" or PABAK which depends solely on P_0 and is given by $PABAK = 2P_0 - 1 = 0.58$. Although some clinicians such as Hoehler (2000) was critical of this index, others such as Mak et al. (2004) found PABAK as a measure of agreement to be important for screening and diagnostic studies in the community setting, when a control for prevalence cannot be easily achieved.

Hirji and Rosove (1990) indicated that a measure of agreement should satisfy the following criteria:

a. In case of perfect agreement, it should have the value 1.
b. In case of perfect disagreement, it should have the value –1.
c. It should adjust for agreement due to chance, and the two raters are independent, it should have the value zero.

One should note that *PABAK* satisfies (a) and (b) but not (c). In the meantime Cohen's kappa satisfies (a) and (c) but not (b). Hirji and Rosove (1990) devised a measure of conditional agreement that satisfies all the three characteristics. They defined

$$1 + \lambda_i = \left(\frac{n_{ii}}{n_{i.}} \right) + \left(\frac{n_{ii}}{n_{.i}} \right).$$

Accordingly, λ_1 is considered an index of agreement for situation in which at least one of the two raters yields a positive result. The parameter λ_2 has similar interpretation. Defining $\lambda = (\lambda_1 + \lambda_2)/2$, satisfies the three conditions of a coefficient of agreement. The notations used here are $n_{11} \equiv a$, $n_{10} \equiv b$, $n_{01} \equiv c$, $n_{00} \equiv d$ for the cell frequencies. The marginal frequencies are $n_{1.} = n_{11} + n_{10}$, $n_{.1} = n_{11} + n_{01}$, $n_{2.} = n_{01} + n_{00}$, $n_{.2} = n_{10} + n_{00}$. The authors argued that the parameter λ can be naturally generalized to the case of more than two categories and it has the ability to gauge conditional agreement. For example, in addition to assessing overall agreement, one may be interested in analyzing concordance within a particular category. To illustrate this situation, where one may be interested in analyzing category specific agreement, assume that $n_{11} \equiv 999$, $n_{10} \equiv 50$, $n_{01} \equiv 50$, $n_{00} \equiv 1$. In this case both kappa = -0.02, and $\lambda = -0.03$ indicate absence of agreement. However, $\lambda_1 = 0.91$ indicates perfect agreement for the positive category and $\lambda_2 = -0.96$ indicates perfect disagreement with respect to the negative category.

In summary, imbalances in the distribution of the marginal totals can sometimes produce two types of paradoxes when the variability of two raters for binary classifications is expressed with the kappa coefficient. Kappa can sometimes be low despite relatively high values of P_0; and will sometimes be increased, rather than decreased by departures from symmetry in the marginal totals. Despite these paradoxical issues, kappa has emerged as a versatile index of nominal scale agreement between two raters.

In face of the above arguments, more recently, Vach (2005) concluded that for binary measurements the dependence of kappa on the observed marginal prevalence is a direct consequence of the definition of kappa and its objective to correct a crude agreement rate relative to the expected agreement by chance. Hence, we concur with him that "it makes no sense to criticize kappa for exactly fulfilling this property." Therefore, one should not regard the dependence of kappa on the marginal prevalence as a drawback.

5.8 Sample Size Requirements

5.8.1 Power Considerations

In assessing interrater reliability, a choice must be made on how to measure the condition under investigation. One of the practical aspects of this decision concerns the relative advantages of measuring the trait on a continuous scale, as in the previous sections, or on dichotomous scale. In many medical screening programs, and studies in social sciences and psychology, it is often feasible to record subject's response on a dichotomous scale.

Donner and Eliasziw (1992) used the goodness-of-fit test procedure to facilitate sample size calculations useful for ensuring enrollment of a sufficient number of subjects in the reliability study. They showed that the number of subjects needed to test $H_0 : \kappa = \kappa_0$ versus $H_0 : \kappa = \kappa_1$ in a 2×2 reliability kappa study is

Number of subjects =

$$A^2 \left\{ \frac{\left[\pi(1-\pi)(\kappa_1 - \kappa_0)\right]^2}{\pi^2 + \pi(1-\pi)\kappa_0} + \frac{2\left[\pi(1-\pi)(\kappa_1 - \kappa_0)\right]^2}{\pi(1-\pi)(1-\kappa_0)} + \frac{\left[\pi(1-\pi)(\kappa_1 - \kappa_0)\right]^2}{(1-\pi)^2 + \pi(1-\pi)\kappa_0} \right\}^{-1},$$

(5.13)

where $A^2 = (z_{1-\alpha/2} + z_{1-\beta})^2$. For different values of α (Type I error) and $1 - \beta$ (power), we list $(z_{1-\alpha/2} + z_{1-\beta})^2$ in Table 5.11.

Example 5.3

Suppose that it is of interest to test $H_0 : \kappa = 0.60$ versus $H_1 : \kappa \neq 0.60$ where $\kappa_0 = 0.60$ corresponds to the value of kappa characterized by Landis and Koch (1977b) as representing substantial agreement. To ensure with 80% probability a significant result at $\alpha = 0.05$ and $\pi = 0.30$ when $\kappa_1 = 0.90$, the required number of subjects from Equation 5.13 is $k = 66$.

In Table 5.12 we present some values of the required number of subjects for different values of κ_0, κ_1, π, $\alpha = 0.05$, and $1 - \beta = 0.80$.

5.8.2 Fixed Length of a Confidence Interval

Similar to the case of continuous measurements, we may base our sample size calculations, on the required length of confidence interval on kappa.

Suppose that an agreement study is to be conducted such that a confidence interval on kappa has a desired length w. Setting

$$w = 2z_{\alpha/2}\sqrt{\text{var}(\hat{\kappa})},$$

where

$$\text{var}(\hat{\kappa}) = \frac{1-\kappa}{k}\left[(1-\kappa)(1-2\kappa) + \frac{\kappa(2-\kappa)}{2\pi(1-\pi)}\right],$$

one can solve for k in terms of κ and π to get the required number of subjects k.

TABLE 5.11

Values of $A^2 = (z_{1-\alpha/2} + z_{1-\beta})^2$

α	$1 - \beta$	A^2
0.05	0.80	7.849
0.05	0.90	10.507
0.01	0.80	11.679
0.01	0.90	14.879

TABLE 5.12

Required Number of Subjects for $\alpha = 0.05$, $\beta = 0.20$

κ_1	π	κ_0				
		0.4	0.6	0.7	0.8	0.9
0.4	0.1		334	121	49	17
	0.3		148	52	21	7
	0.5		126	45	18	6
0.6	0.1	404		1090	195	46
	0.3	190		474	83	20
	0.5	165		400	71	17
0.7	0.1	179	1339		770	103
	0.3	84	595		336	44
	0.5	73	502		282	37
0.8	0.1	101	335	1090		413
	0.3	47	148	474		177
	0.5	41	125	400		149
0.9	0.1	64	149	272	779	
	0.3	30	66	118	336	
	0.5	26	55	100	282	

Replacing κ by a planned value $\bar{\kappa}$ and solving for κ (the required number of subjects), we get

$$k = \frac{4Z_{\alpha/2}^2}{w^2}\left\{(1-\bar{\kappa})\left[(1-\bar{\kappa})(1-2\bar{\kappa}) + \frac{\bar{\kappa}(2-\bar{\kappa})}{2\pi(1-\pi)}\right]\right\}. \qquad (5.14)$$

Example 5.4

Suppose that an interobserver agreement study involving two raters is designed to achieve an agreement coefficient $\kappa = 0.80$. Moreover, it is assumed that the probability of positive rating is 0.30, and the desired width of the confidence interval is $w = 0.20$, and the Type I error rate is set at $\alpha = 0.05$. Then,

$$k = \frac{4(1.64)^2}{(0.2)^2}\left\{(0.2)\left[(0.2)(-0.6) + \frac{(0.8)(1.2)}{2(0.3)(0.7)}\right]\right\} = 117.$$

5.8.3 Efficiency Requirements

The primary focus of this section is on the determination of the optimal allocation of fixed total number of measurements $N = nk$, so that the variance of the estimate of κ is minimized when the response is dichotomous.

Let y_{ij} be the jth rating made on the ith subject, where $y_{ij} = 1$ if the condition is present and 0 otherwise. Landis and Koch (1977), analogous to the continuous case, employed the one-way random effect model:

$$y_{ij} = \mu_i + e_{ij},\tag{5.15}$$

where $\mu_i = \mu + s_i$ for $i = 1, 2, ..., k$, $j = 1, 2, ..., n$. With the usual assumptions for estimation, $\{s_i\}$ are *iid* having mean 0 and variance σ_s^2 and the $\{e_{ij}\}$ are *iid* having mean 0 and variance σ_s^2, and as in the continuous case, the sets $\{s_i\}$ and $\{e_{ij}\}$ are assumed to be mutually independent.

In this context, the standard assumptions for the y_{ij} corresponding to the above ANOVA model are

$$E\left(y_{ij}\right) = \pi = \Pr\left[y_{ij} = 1\right]$$

and that

$$\sigma^2 = \text{var}\left(y_{ij}\right) = \pi\left(1 - \pi\right).\tag{5.16}$$

Moreover, let $\delta = \Pr[y_{ij} = 1, y_{il} = 1] = E(y_{ij}, y_{il})$, then it follows for $j \neq l$ and $i = 1, 2, ..., k$ that

$$\delta = \text{cov}\left(y_{ij}, y_{il}\right) + E\left(y_{ij}\right)E\left(y_{il}\right) = \rho\pi\left(1 - \pi\right) + \pi^2,\tag{5.17}$$

where ρ is the (within-subject) ICC, and as a result, $\rho = (\delta - \pi^2)/[\pi(1 - \pi)]$. In this form it is apparent that the ρ is directly analogous to the components of kappa proposed by Fleiss (1971), and in particular, it represents an agreement measure that allows for the agreement expected by chance, and standardized by the appropriate variances. Following Landis and Koch (1977), let

$$\sigma_s^2 = \rho\pi\left(1 - \pi\right)\tag{5.18}$$

$$\sigma_e^2 = \left(1 - \rho\right)\pi\left(1 - \pi\right)\tag{5.19}$$

be the variance components of y_{ij}. It then follows that the total variance in Equation 5.16 can be written as

$$\sigma^2 = \sigma_s^2 + \sigma_e^2$$

and the (within-subject) ICC ρ can be expressed as

$$\rho = \frac{\sigma_s^2}{\left(\sigma_s^2 + \sigma_e^2\right)}.$$

Mak (1988) established the equivalence between the ICC (ρ) and the coefficient of agreement (κ).

The ANOVA estimate for ρ is given by

$$\rho^* = \frac{MSB - MSW}{MSB + (n-1)MSW},$$

where

$$MSB = \frac{1}{k-1}\left[\frac{\sum_{i=1}^{k} y_i^2}{n} - \frac{\left(\sum_{i=1}^{k} y_i\right)^2}{nk}\right],$$

$$MSW = \frac{1}{k(n-1)}\left[\frac{\sum_{i=1}^{k} y_i - \sum_{i=1}^{k} y_i^2}{n}\right],$$

and $y_i = \sum_{j=1}^{n} y_{ij}$.

The asymptotic variance of ρ^* was given by Mak (1988) but is too complex to allow derivation of an explicit expression for the optimal number of replicates. This difficulty may be overcome by using results from the literature on the estimation for correlated binary response models.

First, we note that the statistic ρ^* depends on the subject's total $y_i = \sum_{j=1}^{n} y_{ij}$, and not on the individual binary responses. Second, Crowder (1978) and Haseman and Kupper (1979) demonstrated the equivalence of the above ANOVA model and the well-known β-binomial model which occurs when conditional on the subject effect μ_i, the subject's total y_i has binomial distribution with conditional mean and variance given, respectively, by

$$E(y_i|\mu_u) = n\mu_i, \quad \text{and} \quad \text{var}(y_i|\mu_i) = n\mu_i(1-\mu_i).$$

With μ_i assumed to follow the β-distribution

$$f(\mu_i) = \frac{\Gamma(a+b)}{\Gamma(a)\Gamma(b)}\mu_i^{a-1}(1-\mu_i)^{b-1}. \tag{5.20}$$

With the appropriate parameterization, $a = \pi(1-\rho)/\rho$, and $b = (1-\pi)(1-\rho)/\rho$. Therefore, ANOVA model and the β-binomial model are virtually indistinguishable (Cox and Snell, 1989). Now, since for the nonnormal case, (as demonstrated in Chapter 2) the optimal number of replicates, under the ANOVA model was shown to be

$$n^* = 1 + \frac{1}{\rho\sqrt{1 + \gamma_s}} \tag{5.21}$$

and since γ_s is the kurtosis of the subject effect distribution, it turns out that, one may use the kurtosis of the β-distribution (the subject-random effect distribution for binary data) to determine the optimal number of replications in the case of dichotomous response.

One can derive the γ_s for β-distribution from the recurrence relation

$$m_1' = \pi,$$

$$m_l' = m_{l-1}' \left[\frac{(l-1)\rho + \pi(1-\rho)}{1 + (l-2)\rho} \right], \quad l = 2, 3, \dots$$

where $m_i' = E\left[\mu_i^l \right]$. Then, $\gamma_s = (m_4/(m_2)^2)$, where

$$m_4 = m_4' - 4m_3'm_1' + 6m_2'\left(m_1' \right)^2 - 3\left(m_1' \right)^4$$

and

$$m_2 = m_2' - \left(m_1' \right)^2$$

Substituting γ_s into Equation 5.21, we obtain

$$n^* = 1 + \pi(1 - \pi) \left[\frac{(1+\rho)(1+2\rho)}{\Psi(\pi, \rho)} \right]^{1/2}, \tag{5.22}$$

where

$$\Psi(\pi, \rho) = \pi\left[\rho + \pi(1-\rho) \right]\left[2\rho + \pi(1-\rho) \right]$$

$$\left[3\rho + \pi(1-\rho) - 4\pi(1+2\rho) \right] + (1+\rho)(1+2\rho)$$

$$\left[6\pi^3 (1-\pi)\rho + 3\pi^4 + \pi^2 (1-\pi)^2 \rho^2 \right].$$

Table 5.13 shows the optimal number of replications (n^*), and the corresponding optimal number of subjects, $k = N/n^*$. In contrast to the continuous

TABLE 5.13

Optimal Allocation Of $N = 60$ for Dichotomous Response

		ρ											
		0.4		0.5		0.6		0.7		0.8		0.9	
		n	k	n	k	n	k	n	k	n	k	n	k
π	0.1	1.81	33	1.64	37	1.53	39	1.46	41	1.40	43	1.36	44
	0.3	2.36	25	2.10	29	1.94	31	1.82	33	1.73	35	1.65	36
	0.5	2.53	24	2.25	27	2.08	29	1.95	31	1.85	32	1.77	34

measurement model, the optimal allocations in the case of dichotomous assessment depend on the mean of the binary response variable π. We also noted that, for fixed N, the allocations are equivalent for π and $1 - \pi$ and therefore, we restricted the values of π in Table 5.13 to $\pi = 0.1$, 0.3, and 0.5.

Remarks

1. When π is small, only as few as two replicates are needed, and a larger number of subjects should be recruited to ensure accurate estimation of ρ.
2. When $\pi = 0.5$, fewer number of subjects should be recruited with no more than three replicates from each subject.
3. In general, as expected, higher values of ρ means that only as few as $n = 2$ replicates are needed, and hence, a larger number of subjects should be recruited. In particular, when $\pi = 0.5$ and $0.6 \leq \rho \leq 0.8$, the optimal allocations for the binary response is close to those required for the normal continuous response.

There are many situations when repeated assessments are made by each of the raters involved in the study.

5.9 Dependent Dichotomous Assessments

5.9.1 Pooled Estimate of Interrater Agreement from Replicated Trials

Oden (1991) proposed a method to estimate a pooled kappa between raters when both raters rate the same set of pairs of eyes. His method assumes that the true left-eye and right-eye kappa values are equal and makes use of the correlated data to estimate confidence intervals for the common kappa.

The pooled kappa estimator is the weighted average of the kappas for the right and left eyes, and is given by

$$
\hat{\kappa}_{\text{Pooled}} = \frac{\left(1 - \sum_{i=1}^{C}\sum_{j=1}^{C} w_{ij}P_{i.}P_{.j}\right)\hat{\kappa}_{\text{right}} + \left(1 - \sum_{i=1}^{C}\sum_{j=1}^{C} w_{ij}q_{i.}q_{.j}\right)\hat{\kappa}_{\text{left}}}{\left(1 - \sum_{i=1}^{C}\sum_{j=1}^{C} w_{ij}P_{i.}P_{.j}\right) + \left(1 - \sum_{i=1}^{C}\sum_{j=1}^{C} w_{ij}q_{i.}q_{.j}\right)}. \tag{5.23}
$$

$i = 1, 2, j = 1, 2$, and c is the number of categories. Moreover, P_{ij} = proportion of patients whose right eye was rated i by rater 1 and j by rater 2.

$$
p_{i.} = p_{i1} + p_{i2} + \cdots + p_{iC}, p_{.j} = p_{1j} + p_{2j} + \cdots + p_{cj}.
$$

q_{ij} = proportion of patients whose left eye was rated i by rater 1 and j by rater 2.

$$
q_{i.} = q_{i1} + q_{12} + \cdots + q_{ic}, q_{.j} = q_{1j} + q_{2j} + \cdots + q_{cj}.
$$

w_{ij} = agreement weight that reflects the degree of agreement between raters 1 and 2 if they use ratings i and j, respectively, for the same eye.

Oden obtained an approximate standard error of the pooled estimator of kappa. The estimator was almost unbiased, and had better performance than either the naive two-eye estimator (which treats the data as a random sample of $2k$ eyes) or the estimator based on either single eye, in terms of correct coverage probability of the 95% confidence interval for the true kappa.

Schouten (1993) presented an alternative approach in this context. He noted that existing formulae for the evaluation of weighted kappa and its standard error can be used if the observed as well as the chance agreement is averaged over the two sets of eyes and then substituted into the expression for kappa estimator in 5.23. This may be explained as follows.

Let each eye be diagnosed normal/or abnormal, and let each patient be classified into one of the following four categories by each rater:

R^+L^+: abnormality is present in both eyes

R^+L^-: abnormality is present in the right-eye but not in the left-eye

R^-L^+: abnormality is present in the left-eye but not in the right-eye

R^-L^-: abnormality is absent in both eyes

Table 5.14 shows the frequency of ratings, together with the weights assigned by Schouten. He used the weights $w_{ij} = 1$ if the raters agree on both

TABLE 5.14

Schouten Frequencies and Weights for the Binocular Data (Bracketed Numbers Are Assigned Weights)

	Rater (2)				
Rater (1)	R^+L^+	R^+L^-	R^-L^+	R^-L^-	Total
R^+L^+	n_{11} (1.0)	n_{12} (0.5)	n_{13} (0.5)	n_{14} (0)	$n_{1.}$
R^+L^-	n_{21} (0.5)	n_{22} (1)	n_{23} (0.0)	n_{24} (0.5)	$n_{2.}$
R^-L^+	n_{31} (0.5)	n_{32} (0.0)	n_{33} (1.0)	n_{34} (0.5)	$n_{3.}$
R^-L^-	n_{41} (0.0)	n_{42} (0.5)	n_{43} (0.5)	n_{44} (1.0)	$n_{4.}$
Total	$n_{.1}$	$n_{.2}$	$n_{.3}$	$n_{.4}$	n

eyes, 0.5 if the raters agree on one eye and disagree on the other, and 0.0 if the raters disagreed on both eyes.

The overall weighted kappa statistics is defined as

$$\hat{\kappa}_w = \frac{P_{ow} - P_{ew}}{1 - P_{ew}},$$

(5.24)

where

$$P_{ow} = \frac{1}{n} \sum_{i=1}^{4} \sum_{j=1}^{4} w_{ij} n_{ij},$$

$$P_{ew} = \frac{1}{n} \sum_{i=1}^{4} \sum_{j=1}^{4} w_{ij} (n_{i.})(n_{.j}).$$

A consistent estimator of the large sample variance of $\hat{\kappa}_w$ was given by Fleiss et al. (1969) as

$$\hat{\text{var}}(\hat{\kappa}_w) =$$

$$\frac{1}{n(1-P_{ew})^2} \left[\sum_{i=1}^{4} \sum_{j=1}^{4} \frac{n_{ij}}{n} \left[w_{ij} - (\bar{w}_{.j} + \bar{w}_{.j})(1 - \hat{\kappa}_w) \right]^2 - \left[\hat{\kappa}_w - P_{ew}(1 - \hat{\kappa}_w) \right]^2 \right].$$

(5.25)

Here,

$$\bar{w}_{i.} = \sum_{j=1}^{4} \frac{n_{.j}}{n} w_{ij},$$

and

$$\bar{w}_{.j} = \sum_{i=1}^{4} \frac{n_{i.}}{n} w_{ij}.$$

An approximate $100(1 - \alpha)$ percent confidence interval is given by

$$\hat{\kappa} \pm z_{\alpha/2}\sqrt{\hat{var}(\hat{\kappa}_w)}.$$

Example 5.4

Oden (1991) provided data concerning the presence/absence of geographic atrophy in the eyes of 840 patients, each eye graded by the same two graders. Table 5.15 gives the observed counts for the cross-classification of patients by their graders 1 and 2 ratings in left and right eyes. We present the data after converting them in a convenient manner to use Schouten's approach.

The bracketed numbers are the assigned weights. Direct computations show that $P_o = 0.98$, $P_e = 0.96$. On substitution in Equation 5.24, we obtain

$$\hat{\kappa}_w = \frac{0.98 - 0.96}{1 - 0.96} = 0.50,$$

which is very close to Oden's pooled estimate of kappa. The $\hat{SE}(\hat{\kappa}_w = 0.110)$, and a 95% confidence interval on the coefficient of agreement is (0.28, 0.77).

5.9.2 Comparing Dependent Interrater Agreements: Probability Models

In the previous section, we demonstrated how to deal with the problem of estimating agreement from dependent or replicated data. The need to compare two or more interrater agreements from the same group of subjects arises in many situations. For example, Bowerman et al. (1990) report on a study in which four readers, two radiologists and two clinical hematologists

TABLE 5.15

Binocular Data

Rater (1)	Rater (2)				Total
	R^+L^+	R^+L^-	R^-L^+	R^-L^-	
R^+L^+	4 (1.0)	0 (0.5)	0 (0.5)	1 (0)	5
R^+L^-	0 (0.5)	5 (1)	0 (0.5)	3 (0.5)	8
R^-L^+	0 (0.5)	0 (0.0)	2 (1.0)	4 (0.5)	6
R^-L^-	2 (0)	9 (0.5)	10 (0.5)	800 (1.0)	821
Total	6	14	12	808	840

independently assessed the vertebral radiographic index (VRI) on 40 radio-graphs from patients with myeloma. One purpose of this investigation was to determine how coefficients measuring interobserver agreement varied according to expertise in radiologic diagnosis. This lead to the comparison of two such coefficients, one for radiologists and one for the nonradiologists, with each coefficient computed from data collected on the same set of 40 subjects.

Similar problems arise when it is of interest to compare coefficients of intraobserver variability. For example, Baker et al. (1991) report on a study in which each of two pathologists assessed 27 patients with respect to the presence or absence of dysplasia. Each assessment was performed in duplicate, providing an opportunity to investigate whether the two pathologists showed comparable levels of within-observed reproducibility. This comparison again suggests a test of equality between two dependent kappa statistics, where each statistic may be regarded as an index of agreement.

Recall that for the case of a continuous outcome variable (Chapter 2), the ICC is frequently used as a measure of interobserver reliability. Test for comparing two dependent intraclass correlations have been introduced by Alsawalmeh and Feldt (1994) and was discussed in Chapter 2.

In this section we focus on the problem of comparing two intraclass kappa statistics as computed over the same sample of subjects, that is, we relax the assumption of independent samples. Relatively little research has been reported on this topic, and many such studies were concerned with reporting descriptive comparisons only. This problem was noted by McKenzie et al. (1996) who remarked that "methods for the comparison of correlated kappa coefficients obtained from the same sample do not appear in the literature." These authors developed a testing procedure which uses computationally intensive resampling methods. Attention was limited to tests of pairwise equality among kappa coefficients in studies where each subject is assessed by three raters. Other research on this problem has been reported by Williamson and Manatunga (1997). They described an estimating equations approach for the analysis of ordinal data with the underlying assumption of a latent bivariate normal variable, which is computationally intensive as well.

For the models discussed in this section we assume that each of N subjects is rated under two settings, where each setting involves each of two observers assigning a binary rating to each subjects. The probability model developed by Donner et al. (2000) and Shoukri and Donner (2001) can be characterized by the parameters $\kappa_j(j = 1, 2)$ and κ_b, where κ_j measures the level of interrater agreement under setting j and κ_b measures the expected level of interrater agreement between any two ratings selected on the same subject from the different settings. The focus in this section is on the estimation of $\kappa_j(j = 1, 2)$, and κ_b, and testing the hypothesis $H_o: \kappa_1 = \kappa_2$, where κ_b is primarily regarded as a nuisance parameter.

Let $Y_{ijk} = 1(0)$ denote the binary assignment of the ith subject under the jth setting for rater kth rater as a success (failure), $i = 1, 2, ..., N, j = 1, 2; k = 1, 2$.

Furthermore, let π denote the marginal probability that an observation is recorded as a success across all subjects in the population, and let P_i denote the probability that an observation for the ith subject is recorded as a success as averaged over both settings. A mechanism for introducing the correlation between ratings obtained in different settings is to assume that conditional on π, the distribution of P_i among all subjects is modeled as a β-distribution with parameters a and b, that is $P_i | \pi \sim f(P_i) = \beta(a, b)$, with $\pi = a/(a + b)$. Furthermore, let P_{ij} denote the probability that an observation is recorded as a success to the ith subject under the jth setting. Following Rosner (1989, 1992), we introduce the between-setting correlation by assuming that, conditional on P_i, the distribution of P_{ij} is modeled as $f(P_{ij}) \sim \beta - (\lambda_j P_i, \lambda_j (1 - P_i))$.

Let $\kappa_{cj} = Corr(Y_{ij1}, Y_{ij2} | P_i) = (1 + \lambda_j)^{-1}$. The parameter κ_{cj} may be interpreted as the conditional within-setting correlation. Let $\kappa_b = Corr(Y_{ijk}, Y_{ij'k}) = (1 + a + b)^{-1}, j \neq j'$, denote the between-setting correlation.

Table 5.16 depicts the layout for this type of dependent data.

Donner et al. (2000) showed that these assumptions lead to the following probabilities for the joint distribution of the ratings (Y_{ij1}, Y_{ij2}) taken under the jth setting.

$$P_o(\kappa_j) = P_r(X_{ij1} = 0, X_{ij2} = 0)$$

$$= (1 - \pi)^2 + \kappa_j \pi (1 - \pi),$$

$$P_1(\kappa_j) = P_r(y_{ij1} = 1, y_{ij2} = 0 \quad \text{or} \quad y_{ij1} = 0, y_{ij2} = $$

$$= 2\pi(1 - \pi)(1 - \kappa_j),$$

$$P_2(\kappa_j) = P(y_{ij1} = 1, y_{ij2} = 1)$$

$$= \pi^2 + \kappa_j \pi (1 - \pi),$$

TABLE 5.16

Data Layout for N Subjects Evaluated by Two Raters under Two Settings

Setting		Subject				
		1	2	3	...	N
(1)	Rater (1)	y_{111}	y_{211}			y_{N11}
	Rater (2)	y_{112}	y_{212}			y_{N12}
(2)	Rater (1)	y_{121}	y_{221}			y_{N21}
	Rater (2)	y_{122}	y_{222}			y_{N22}

TABLE 5.17

Marginal Ratings Frequencies of the N Subjects
under Each of the Two Settings

Category	Probability	Ratings	Frequency of Subjects	
			Setting 1	Setting 2
0	$P_0(\kappa_j)$	(0, 0)	$n_{.00}$	$n_{0.0}$
1	$P_1(\kappa_j)$	(1, 0) or (0, 1)	$n_{.01}$	$n_{1.0}$
2	$P_2(\kappa_j)$	(1,1)	$n_{.02}$	$n_{2.0}$

where

$$\kappa_j = \kappa_b + \kappa_{cj}(1 - \kappa_b)j = 1,2. \tag{5.26}$$

Therefore, κ_j in $P_\alpha(\kappa_j)$, $\alpha = 1, 2, 3$ is the weighted sum of κ_{cj}, and its maximum value 1, with respective weights κ_b and $(1 - \kappa_b)$.

The multivariate distribution of the binary scores has therefore $2^4 = 16$ probabilities, corresponding to the possible binary assignments on a given subject. The details of the derivations and the 16 probabilities are given in Donner et al. (2000) (see Table 5.17).

Since we assumed that the two ratings by one rater are exchangeable, the 2^4 probabilities have a parametric structure with symmetries that allow collapsing to a 3^2 as shown in Table 5.18.

The cell probabilities are given in Appendix 5.1 at the end of this section. In this special case $\kappa_b = 0$, the MLE of κ_1 and κ_2 are given by

$$\hat{\kappa}_1 = 1 - \frac{n_{0.1}}{2N\hat{\pi}(1 - \hat{\pi})}, \tag{5.27a}$$

TABLE 5.18

Joint Distribution of Sum of First Setting Scores (X_{i1}) and Sum of Second Setting Scores (X_{i2}). Bracketed Numbers Are Observed Frequencies

		X_{i1}			
		0	1	2	Total
X_{i2}	0	θ_{00}	θ_{01}	θ_{02}	$\theta_{0.0} = P_1(\kappa_2)$
		(n_{00})	(n_{01})	(n_{02})	
	1	θ_{10}	θ_{11}	θ_{12}	$\theta_{1.0} = P_2(\kappa_2)$
		(n_{10})	(n_{11})	(n_{12})	
		θ_{20}	θ_{21}	θ_{22}	$\theta_{2.0} = P_3(\kappa_2)$
	2	(n_{20})	(n_{21})	(n_{22})	
Total		$\theta_{.00} = P_1(\kappa_1)$	$\theta_{.01} = P_2(\kappa_1)$	$\theta_{.02} = P_3(\kappa_1)$	1

$$\hat{\kappa}_2 = 1 - \frac{n_{1.0}}{2N\hat{\pi}(1-\hat{\pi})}, \tag{5.27b}$$

where

$$\hat{\pi} = \frac{1}{4n}\left[n_{1.0} + n_{0.1} + 2(n_{2.0} + n_{0.2})\right]. \tag{5.28}$$

Shoukri and Donner (2001) showed that a moment estimator of κ_b is

$$\hat{\kappa}_b = \frac{(n_{11} + 2n_{12} + 2n_{21} + 4n_{22}) - 4N\hat{\pi}^2}{4N\hat{\pi}(1-\hat{\pi})}. \tag{5.29}$$

A consistent estimator of the variance of $\hat{\kappa}_b$ is obtained on replacing θ_{ij} by n_{ij}/N, $\theta_{i\cdot0}$ by $n_{i\cdot0}/N$, $\theta_{0.j}$ by $n_{0.j}/N$ (see Appendix 5.2).

Under $H_o : \kappa_1 = \kappa_2$, a suggested overall measure of agreement may be taken as

$$\kappa = \frac{(\hat{\kappa}_1 + \hat{\kappa}_2)}{2}.$$

Donner et al. (2000) investigated the behavior of several statistics on the above hypothesis through an extensive Monte Carlo simulation. These statistics are

1. The goodness-of-fit statistics for testing $H_o : \kappa_1 = \kappa_2$ assuming independence of settings (i.e., $\kappa_b = 0$). This is given by

$$X_G^2 = \sum_{j=0}^{2}\frac{\left[n_{.0j} - N\hat{P}_j(\hat{\kappa})\right]^2}{N\hat{P}_j(\hat{\kappa})} + \sum_{j=0}^{2}\frac{\left[n_{j.0} - N\hat{P}_j(\hat{\kappa})\right]^2}{N\hat{P}_j(\hat{\kappa})} \tag{5.30}$$

   where we obtain $\hat{P}_j(\hat{\kappa})$ replacing π by $\hat{\pi}$ and $\hat{\kappa}_j$ by $\hat{\kappa}$ in $P_j(\hat{\kappa}_l)$ ($l = 1$, $2; j = 0, 1, 2$). Under H_o, X_G^2 follows an approximate chi-square distribution with one degree of freedom.

2. It is expected, when ignoring the dependence between the two settings that X_G^2 would yield type I error lower than nominal. However, the statistic X_G^2 may be extended to the case of dependent settings by adjusting its value to account for $\text{Corr}(\hat{\kappa}_1, \hat{\kappa}_2)$.

   Donner et al. (2000) proposed adjusting X_G^2 so that

$$X_{GD}^2 = \frac{X_G^2}{\left[1 - \text{Corr}(\hat{\kappa}_1, \hat{\kappa}_2)\right]} \tag{5.31}$$

is referred to the table of chi-square with one degree of freedom. Here, $Corr(\hat{\kappa}_1, \hat{\kappa}_2)$ is the estimated correlation between $\hat{\kappa}_1$ and $\hat{\kappa}_2$ (see Appendix 5.2).

3. Letting $\widehat{cov}(\hat{\kappa}_1, \hat{\kappa}_2)$ denote the estimated covariance between $\hat{\kappa}_1$ and $\hat{\kappa}_2$ as given in Appendix 5.2, an alternative test procedure can be constructed by computing the Wald statistic:

$$Z_{VD} = \frac{\hat{\kappa}_1 - \hat{\kappa}_2}{\left[\widehat{var}(\hat{\kappa}_1) + \widehat{var}(\hat{\kappa}_2) - 2\widehat{var}(\hat{\kappa}_1, \hat{\kappa}_2) \right]^{1/2}} \tag{5.32}$$

and referring Z_{VD} to tables of the standard normal distribution. Here

$$\widehat{var}(\hat{\kappa}_j) = \frac{1 - \hat{\kappa}_j}{N} \left[(1 - \hat{\kappa}_j)(1 - 2\hat{\kappa}_j) + \frac{\hat{\kappa}_j(2 - \hat{\kappa}_j)}{2\hat{\pi}(1 - \hat{\pi})} \right].$$

4. Setting $\widehat{cov}(\hat{\kappa}_1, \hat{\kappa}_2) = 0$ in Equation 5.32 we denote the resulting test statistic by Z_V.

Donner et al. (2000) conducted Monte Carlo simulations for $N = 200, 100, 50, 25$ to test $H_o : \kappa_1 = \kappa_2$ at $\alpha = 0.05$, and various values of κ, π, and κ_b. The main conclusions from their results were that the unadjusted test X_G^2, Z_V are overly conservative (using an arbitrary definition in which the empirical type I error is observed to be less than 0.03) only when κ_b is equal in magnitude to κ, that is, when the "between-setting correlation" is equal to the null value of the "within-setting correlation." Otherwise, the observed type I errors for both X_G^2 and Z_V tend to be close to nominal, not withstanding the lack of independence between $\hat{\kappa}_1$ and $\hat{\kappa}_2$. In fact if $\hat{\kappa}_b$ is much less than κ, then adjustment of either Z_V or X_G^2 can lead to inflated type I errors, particularly in small samples, where the empirically estimated adjustment factor lacks stability. The unadjusted statistic Z_V is also frequently too liberal in small samples ($N = 50$ or 25), particularly when $\pi < 0.3$.

Example 5.5

Baker et al. (1991) reported on a study in which each of two pathologists assessed 27 patients with respect to the presence or absence of dysplasia. Each assessment was performed in duplicate. One of the objectives was to investigate whether the two pathologists showed comparable levels of within rater reproducibility. The data are presented in Table 5.19.

TABLE 5.19

Ratings of 27 Patients Obtained from Duplicate
Assessments by Each of the Two Pathologists

Pathologist 2	Pathologist 1			
	(0, 0)	(0, 1) or (1, 0)	(1, 1)	Total
(0, 0)	9	5	6	20
(0,1) or (1, 0)	0	1	0	1
(1,1)	1	1	4	6
Total	10	7	10	27

$\pi = 0.37$, $\hat{\kappa}_1 = 0.48$, $\hat{\kappa}_2 = 0.90$, and $\hat{\kappa}_b = 0.17$. The values of the adjusted test statistics are given by $X_G^2 = 11.12(p = 0.00085)$, $Z_V = 2.14(p = 0.032)$. The values of the adjusted test statistics are given by $X_{GD}^2 = 14.4(p = 0.00015)$, and $Z_{VD} = 2.5(p = 0.012)$. The similarity in the conclusion ($H_o : \kappa_1 = \kappa_2$ is not supported by the data) can be attributed in part to the relatively small value of $\hat{\kappa}_b$.

5.10 Adjusting for Covariates

It is quite likely in many interobserver agreement studies, that the marginal probability of classification for a particular subject may depend on one or more subject-specific covariates. For example, as we have demonstrated in this chapter, the level of agreement between MRI and ultrasound may depend on magnitude of the tumor, which in turn affects the probability of classification positive. Another example (see Barlow, 1996), a radiologist may be more likely to classify a mammographic abnormality as breast cancer when there is a known family history of breast and the patient is elderly, because both age and family history are known risk factors for this disease. Similarly, if we measure the agreement between dichotomous outcomes on the same patient, fixed covariates may influence the probability of the outcome. Barlow et al. (1991) proposed a stratified kappa that can adjust for categorical confounders. It is assumed that the underlying kappa is common across strata, but the marginal probability of classification may differ across table. Failure to account for these confounders (i.e., collapsing across strata) may lead to inflated estimates of agreement (Barlow et al., 1991). It should be emphasized that as the number of confounders becomes large, then the stratified kappa may be based on few observations in each table and would have poor precision. Therefore, it is more efficient to directly adjust for the subject-specific covariates that may influence the raters in their classification of the subject.

In this section, we discuss several approaches for modeling agreement as function of several covariates. We begin by a simple approach proposed by Coughlin et al. (1992).

5.10.1 Logistic Modeling

Additive linear models for categorical data that allow for the testing of hypotheses concerning interobserver agreement and the estimation of coefficient of agreement such as kappa have been described by Landis and Koch (1977). More recently, hierarchical log-linear models have been proposed for the description of interrater agreement involving nominal or ordinal categorical variables (Tanner and Young, 1985; Agresti, 1988).

Recently, Coughlin et al. (1992) suggested a logistic model approach for modeling crude agreement which is adjusted for covariates included in the regression equation. If each of k subjects is assigned independently by two raters to one of c categories, then the cell frequencies (n_{ii}) along the main diagonal of the two-way contingency table represent the agreement between the raters. This may be summarized in terms of the overall proportion agreement P_o which is estimated by

$$P_0 = \frac{1}{k} \sum_{i=1}^{c} n_{ii}.$$

Coughlin et al. (1992) argued that this summary measure is too crude, as it may mask important variation in agreement across subgroups of the sample. The logistic model may be used to estimate the proportion agreement for particular subgroups by defining the dependent variable (y) to be equal to 1 if the two raters agree or else 0. The covariates included in the regression equation define the subgroups of interest and any factors to be adjusted for. Thus, the log-odds of agreement is modeled as a linear function of p explanatory variables $(1, x_1, x_2, ..., x_p)$. The proportion agreement may be estimated for a particular subgroup using the equation

$$E\left(P_o | x_1, ..., x_P\right) = \frac{e^{\eta}}{1 + e^{\eta}}, \tag{5.33}$$

where, $\eta = \beta_o + \sum_{r=1}^{p} \beta_r x_r$.

The variance of the logit of the proportion agreement at a particular covariate level, adjusting for the effects of the other covariates, may be calculated using the following methods of asymptotic variance estimation (see Hosmer and Lemeshow, 1989):

$$\sigma^2 = \text{var}\left[\log P_o/1 - P_o\right] = \text{var}\left(\beta_o\right) + \sum_{r=1}^{p} x_r^2 \, \text{var}\left(\hat{\beta}_r\right)$$

$$+ 2\sum_{r=1}^{P} x_r \text{cor}\left(\hat{\beta}_o, \hat{\beta}_r\right) + \sum\sum_{r\neq l} x_r x_l \, \text{cov}\left(\hat{\beta}_r, \hat{\beta}_l\right).$$

To determine the statistical significance of differences in agreement across subgroups, hierarchal logistic models may be fitted which are identical except for covariate(s) representing the subgroups of interest. Levels of statistical significance may then be obtained using the log-LRT. The same approach may be used to obtain estimates adjusted for one or more confounding variables by substituting the mean value of the respective covariate in Equation 5.33.

These methods may be readily extended to examine the degree of agreement, following general developments in logistic modeling. For example, consider the 3×3 table of classification by two raters. The dependent variable Y may be assigned the value of 2 if the raters agree exactly, 1 if their ratings differ by exactly one category, and 0 if they disagree by two categories; that is,

$$y_{ij} = 2 \quad \text{if}\,|i - j| = 0,$$

$$y_{ij} = 1 \quad \text{if}\,|i - j| = 1,$$

$$y_{ij} = 0 \quad \text{if}\,|i - j| = 2.$$

For $i,j = 1, 2, \ldots, c$.

For $j = 1, 2$, the expected proportion of agreement may be represented as

$$E\left(P_{oj}\middle| X_1, \ldots, X_p\right) = \frac{e^{\eta_j}}{1 + e^{\eta_1} + e^{\eta_2}}, \quad \text{where } \eta_j = \alpha_j + \sum_{r=1}^{p} \beta_{jr} x_r, \quad j = 1, 2.$$

As before, the log odds of degree of agreement are modeled as a linear function of explanatory variables.

5.10.2 Likelihood-Based Approach

For the case of two raters and dichotomous ratings, Shoukri and Mian (1996) used a more general set-up. When the two binary ratings on subject i depend on subject and/or raters' effects, we may define a covariate vector z'_{ij} so that

$$\log it\left(\pi_{ij}\right) = z'_{ij}\beta.$$

Let (y_{i1}, y_{i2}), $i = 1, 2, ..., k$ denote a random sample of k pairs of correlated binary responses whose joint distributions is $f(y_{i1}, y_{i2})$, where

$$f(y_{i1}, y_{i2}) = \left[\pi_{i1}\pi_{i2} + \frac{\kappa}{2}(\pi_{i1}\pi'_{i2} + \pi_{i2}\pi'_{i1}) \right]^{y_{i1}y_{i2}}$$

$$\left[\pi_{i1}\pi'_{i2} - \frac{\kappa}{2}(\pi_{i1}\pi'_{i2} + \pi_{i2}\pi'_{i1}) \right]^{y_{i1}(1-y_{i2})}$$

$$\left[\pi_{i2}\pi'_{i1} - \frac{\kappa}{2}(\pi_{i1}\pi'_{i2} + \pi_{i2}\pi'_{i1}) \right]^{y_{i2}(1-y_{i1})}$$

$$\left[\pi_{i1}\pi'_{i2} + \frac{\kappa}{2}(\pi_{i1}\pi'_{i2} + \pi_{i2}\pi'_{i1}) \right]^{(1-y_{i1})(1-y_{i2})}.$$

Here $\pi'_{ij} = 1 - \pi_{ij}$ $i = 1, 2, ..., k$; $j = 1, 2$, and k is the number of subjects. The log-likelihood function is given by

$$L(\kappa, \beta) = \sum_{i=1}^{k} \log f(y_{i1}, y_{i2})$$

is maximized with respect to κ and β. Shoukri and Mian (1996) obtained the large sample variances and covariances of $\hat{\kappa}$ and $\hat{\beta}$ by inverting the Fisher's information matrix.

Example 5.6

Shoukri and Mian (1996) analyzed data, reported by Hui and Walter (1980) on a new test (named rater 2) when evaluated against the standard Mantoux (rater 1) for the detection of tuberculosis. Both are skin tests applied to the arms. After 48 h the presence of an induration larger than a fixed size constitutes a positive result. Data for population 1 came from a study conducted in a southern U.S. school district. Under the same rating protocol the second study was conducted at the Missouri State Sanatorium. The data appears in Table 5.20. We define the following dummy variables

$$Z_1 = \begin{pmatrix} 1 & \text{if rater 2} \\ 0 & \text{if rater 1}' \end{pmatrix}$$

$$Z_2 = \begin{pmatrix} 1 & \text{if population 1} \\ 0 & \text{if population 2} \end{pmatrix}$$

TABLE 5.20

Hui and Walter's Data

Response	Populations	
(Test 1, Test 2)	(1)	(2)
(Y, Y)	14	887
(Y, N)	4	31
(N, Y)	9	37
(N, N)	52	367

Note: Y = positive, N = negative.

The proposed model is

$$\log it\left(\pi_{ij}\right) = \beta_0 + \beta_1 z_{ij1} + \beta_2 z_{ij2}i = 1, 2, \ldots, k, \quad j = 1, 2$$

The results of the data analysis are summarized in Table 5.21.

Note that, a nonsignificant β_1 means that a marginal homogeneity assumption would be justifiable and significant β_2 indicates that the level of agreement between the two tests is not constant across populations (strata). Moreover, if we construct a pooled estimate of κ similar to that proposed by Barlow et al. (1991):

$$\hat{\kappa}_{pooled} = \frac{\sum \hat{w}_i \hat{\kappa}_i}{\sum \hat{w}_i}$$

$$\hat{SE}\left(\hat{\kappa}_{pooled}\right) = \left(\sum \hat{w}_i\right)^{-1/2}.$$

The $\hat{\kappa}_i$ is the estimate of agreement between the two tests obtained from the ith table, and $\hat{W}_i = (var(\hat{\kappa}_i))^{-1}$, $\hat{\kappa}_1 = 0.547(0.104)$, $\hat{\kappa}_2 = 0.878(0.0144)$. Therefore, $\hat{\kappa}_{pooled} = 0.873$, and $\widehat{SE}\hat{\kappa}_{pooled} = \sqrt{0.00021} = 0.0143$. These values are very close to the MLE in Table 5.21.

TABLE 5.21

Maximum Likelihood Estimates for the Data in Table 5.20

Parameter	Estimate	Standard Error	Z-Value	P-Value
β_0	0.8547	0.0596	14.34	0.000
β_1	−0.0366	0.0302	−1.21	0.226
β_2	−3.9501	0.2137	−18.48	0.000
κ	0.8651	0.0148	58.45	0.000

There has been similar likelihood-based approach to construct inference on agreement. Fitzmaurice et al. (1995) constructed a likelihood-based approach, but measuring the degree of interrater agreement was by using odds ratio. These likelihood-based approaches have been limited to the special case where there are no more than two raters per subject.

5.10.3 Estimating Equations Approach

Molenberghs et al. (1996) described how estimating equations approach of Liang and Zeger (1986) can be used to construct inferences on kappa when there may be more than two raters per subject. Klar et al. (2000) proposed using estimating equations to identify covariates which are associated with the marginal probability of classification by each rater and to identify covariates associated with kappa. They considered a logistic regression model to identify covariates associated with the marginal probability of classification by each rater. A second model, based on Fisher's Z-transformation, is used by them to identify covariates associated with kappa. A key advantage is that an arbitrary and variable number of raters per subject may be used to construct these models yet omitting the need for any stringent parametric assumptions.

5.11 Simultaneous Assessment of Two Binary Traits by Two Raters

Procedures for testing the equality of two kappa statistics measuring interobserver agreement with respect to a single binary trait for the same sample of subjects tested under two experimental settings have been considered by Donner et al. (2000). A limitation of their model is the assumption that a common prevalence parameter characterizes the two settings. However, in comparing levels of interobserver agreement across different traits measured on the same subject, this assumption will generally not be satisfied.

In the following sections, we develop a model and accompanying inference procedure that do not require the assumption of a common prevalence. We then present an example dealing with the agreement between two rating instruments that are frequently used by clinicians in the diagnosis of psychiatric disorders.

5.11.1 Model-Based Approach

We assume that each of N subjects is evaluated independently for the presence/absence of two binary traits by each of two observers. Following the approach of Rosner [5, 6], let $X_{ijh} = 1(0)$ denote the presence (absence) of the jth trait in the ith subject, recorded by the hth observer, $i = 1, 2, ..., N, j = 1, 2,$ and $h = 1, 2$. It is further assumed that for trait j, the N observations, X_{ijh}, are a random sample from a potentially infinite number of independent observations with probability P_{ijh}, where $P_{ijh} = \Pr [X_{ijh} = 1 \mid i, h]$. Defining $E_i (P_{ijh}) = P_{jh}$ as the average of the P_{ijh} over the population of subjects, the average of the P_{jh} over the population of raters is then given by $E_h (P_{jh}) = P_j$.

Letting $X_{ij} = X_{ij1} + X_{ij2}, j = 1, 2$, we may now express

$$\Pr\left(X_{ij} = 0 \middle| P_j\right) = \Pr\left(X_{ij1} = 0, X_{ij2} = 0\right) = \left(1 - P_j\right)^2 + \rho_j P_j \left(1 - P_j\right), \quad (5.33a)$$

$$\Pr\left(X_{ij} = 1 \middle| P_j\right) = \Pr\left(X_{ij1} = 1, X_{ij2} = 0\right) + \Pr\left(X_{ij1} = 0, X_{ij2} = 1\right)$$

$$= 2P_j \left(1 - P_j\right)\left(1 - \rho_j\right), \quad (5.33b)$$

$$\Pr\left(X_{ij} = 2 \middle| P_j\right) = \Pr\left(X_{ij1} = 1, X_{ij2} = 1\right) = P_j^2 + \rho_j P_j \left(1 - P_j\right). \quad (5.33c)$$

Moreover, it can be easily shown that

$$E\left(X_{ij} \middle| P_j\right) = 2P_j \quad \text{and} \quad \text{var}\left(X_{ij} \middle| P_j\right) = 2P_j \left(1 - P_j\right)\left(1 + \rho_j\right).$$

As shown by Mak (1988), the β-binomial distribution with parameters (a_j, b_j) is a special case of Equations 5.33a through 5.33c, with $a_j = (1 - \rho_j)P_j/\rho_j$ and $b_j = (1 - \rho_j)P_j/\rho_j$. Under the assumption of exchangeability, and conditional on the prevalence P_j, the parameter ρ_j may now be interpreted as the interobserver agreement coefficient between the two raters for trait j.

If X_{i1} and X_{i2} are uncorrelated, then inferences conducted under the two separate common correlation models (CCMs) would be fully informative. Since this is an unrealistic assumption in the present context, our aim here is to construct a bivariate distribution for the two traits with marginals given by the CCM.

From Lancaster (1969), the canonical expansion for a continuous distributed bivariate random vector is given by:

$$g\left(p_1, p_2\right) = f(p_1)f\left(p_2\right)\left[1 + \rho_b \frac{\left(p_1 - E(p_1)\right)\left(p_2 - E(p_2)\right)}{\sqrt{\text{var}\left(p_1\right)\text{var}\left(p_2\right)}}\right], \quad 0 \le p_j \le 1.$$

Here, we conveniently set

$$f\left(p_j\right) = \beta\left(\frac{\left(1 - \rho_b\right)\pi_j}{\rho_b}, \frac{\left(1 - \rho_b\right)\left(1 - \pi_j\right)}{\rho_b}\right), \quad j = 1, 2.$$

Then, $F(P_{ij}) = \pi_j$, $\text{var}(P_{ij}) = \rho_{ij}$ π_j $(1 - \pi_j)$ and $E(P_1 P_2) = \pi_1 \pi_2 +$ $\rho_b \sqrt{\pi_1 \pi_2 (1 - \pi_1)(1 - \pi_2)}$, where ρ_b denotes the correlation between P_1 and P_2. Let

$$\vartheta_{lm} = \Pr(X_{i1} = l, X_{i2} = m) = \int_0^1 \int_0^1 \Pr(X_{i1} = l|P_1)\Pr(X_{i2} = m|P_2) g(P_1, P_2) dP_1 dP_2.$$

We show in Appendix 5.3 that the elements of the joint distribution of (X_{i1}, X_{i2}), given by the nine probabilities $\vartheta_{lm}, l, m = 0,1,2$ outlined in Table 5.22, can be written as linear combinations of the higher-order bivariate noncentral moments, given by $\psi(l,m) = E(P_1^l P_2^m), l = 0, 1, 2; m = 0, 1, 2$ and the correlation parameters ρ_1 and ρ_2.

It can now easily be shown that

$$E(P_{i1}^l P_{i2}^m) \equiv \psi(l,m) = \psi_1(l)\psi_2(m)$$

$$+ \frac{\rho_b}{\sigma_1 \sigma_2}\left[\psi_1(l+1) - \pi_1\psi_1(l)\right]\left[\psi_2(m+1) - \pi_2\psi_2(m)\right],$$

where

$$\sigma_j^2 = \rho_b \pi_j (1 - \pi_j), \psi_j(l) = E(P_j^l) = \frac{\Gamma(l + a\pi_j)}{\Gamma(a\pi_j)} \cdot \frac{\Gamma(a)}{\Gamma(l + a)} \quad \text{and} \quad a = \frac{1 - \rho_b}{\rho_b}.$$

We also note that to ensure that the probabilities ϑ_{lm} are nonnegative, ρ_b must satisfy the inequality

$$\frac{1}{\rho_1 \rho_2}\max\left[-\frac{1}{\pi_1 \pi_2}, -\frac{1}{(1 - \pi_1)(1 - \pi_2)}\right] \le \rho_b \le \frac{1}{\rho_1 \rho_2}\min\left[\frac{1}{\pi_1(1 - \pi_2)}, \frac{1}{\pi_2(1 - \pi_1)}\right].$$

TABLE 5.22

Joint Probability Distribution of (X_{i1}, X_{i2})

Trait 2 X_2	Trait 1 X_1			Total
	0	1	2	
0	ϑ_{00}	ϑ_{01}	ϑ_{02}	$P_{0.}$
1	ϑ_{10}	ϑ_{11}	ϑ_{12}	$P_{1..}$
2	ϑ_{20}	ϑ_{21}	ϑ_{22}	$P_{2.}$
Total	$P_{.0}$	$P_{.1}$	$P_{.2}$	1

To estimate the model parameters, we assume that the N subjects can be classified according to the presence or absence of one of each trait as given in Table 5.22.

Dropping the subscript, we can now show that the marginal distribution of X_1 is given by

$$P_{.0} = \left(1 - \pi_1\right)^2 + k_1\pi_1\left(1 - \pi_1\right),$$

$$P_{.1} = 2\pi_1\left(1 - \pi_1\right)\left(1 - k_1\right),$$

$$P_{.2} = \pi_1^2 + k_1\pi_1\left(1 - \pi_2\right),$$

where $k_1 = \rho_b + \rho_1\left(1 - \rho_b\right)$. Moreover, X_2 has an identical marginal distribution with probabilities $P_{.0}$, $P_{.1}$, and $P_{.2}$, when π_1, k_1 is replaced by π_2, k_2.

Thus, the bivariate model giving the probabilities $P_{.l}$, $P_{.m}$, $l,m = 0, 1, 2$ is identical to the two CCMs separately if $k_j = \rho_b + \rho_j\left(1 - \rho_b\right)$ replaces ρ_j. However, the bivariate model is more flexible as it has the ability to test whether the level of interobserver agreement is the same across the two traits. It may also be regarded as a generalization of a model proposed by Shoukri and Donner (2001), who investigated the gain in precision obtained by increasing the number of measurements on a single binary outcome measure from 1 to 2.

One should also note that the marginal variance of X_{ij} is given by

$$\text{var}\left(X_{ij}\right) = 2\pi_j\left(1 - \pi_j\right)\left(1 - \rho_b\right)\left[1 + \rho_j + \frac{2\rho_b}{\left(1 - \rho_b\right)}\right].$$

Moreover, the correlation between (X_{i1}, X_{i2}) can be shown to be given by

$$\text{corr}\left(X_{i1}, X_{i2}\right) = \frac{2}{\sqrt{\left(1 + k_1\right)\left(1 + k_2\right)}}.$$

Maximum likelihood estimation is frequently used to estimate the parameters in the vector $\varpi = \left(\pi_1, \pi_2, k_1, k_2, \rho_b\right)'$. Here, we use an alternative two-step approach to create pseudo-MLE, as introduced by Gong and Samaniego (1981) (see Table 5.23).

In the first step, parameters (π_1, π_2, k_1, k_2) are estimated by maximum likelihood using the observed marginal distributions of X_1 and X_2. From Bloch and Kraemer (1989) the MLE are given as

$$\hat{\pi}_1 = \frac{1}{2N}\left(n_{.1} + 2n_{.2}\right), \quad \hat{\pi}_2 = \frac{1}{2N}\left(n_{1.} + 2n_{2.}\right), \quad \hat{k}_1 = 1 - \frac{n_{.1}}{2N\hat{\pi}_1\left(1 - \hat{\pi}_1\right)},$$

TABLE 5.23

Joint Frequencies of Two Traits for a Sample of N Patients

Observer 2	Observer 1			Total
	0	1	2	
0	n_{00}	n_{01}	n_{02}	$n_{0.}$
1	n_{10}	n_{11}	n_{12}	$n_{1.}$
2	n_{20}	n_{21}	n_{22}	$n_{2.}$
Total	$n_{.0}$	$n_{.1}$	$n_{.2}$	N

$$\hat{k}_2 = 1 - \frac{n_{1.}}{2N\hat{\pi}_2(1 - \hat{\pi}_2)}.$$

In the second step, a moment estimator of ρ_b may be obtained as

$$\hat{\rho}_b = \frac{m_{11} - 4\hat{\pi}_1\hat{\pi}_2}{4\left[\hat{\pi}_1\hat{\pi}_2(1 - \hat{\pi}_1)(1 - \hat{\pi}_2)\right]^{1/2}},$$

where

$$m_{11} = N^{-1}(n_{11} + 2n_{12} + 2n_{21} + 4n_{22}).$$

We have already stated that Bloch and Kraemer (1989) gave the following expressions:

$$\text{var}(\hat{\pi}_j) = \frac{\pi_j(1 - \pi_j)(1 + k_j)}{2N}$$

and

$$\text{var}(\hat{k}_j) = \frac{1 - k_j}{N}\left[(1 - k_j)(1 - 2k_j) + \frac{k_j(2 - k_j)}{2\pi_j(1 - \pi_j)}\right].$$

Furthermore, Shoukri and Donner (2009) showed that

$$\text{cov}(\hat{\pi}_1, \hat{\pi}_2) = \frac{4\rho_b\tau}{N},$$

where

$$\tau^2 = \pi_1\pi_2(1 - \pi_1)(1 - \pi_2).$$

The full variance–covariance matrix of the estimated parameter vector $\hat{\omega} = (\hat{\pi}_1, \hat{\pi}_2, \hat{k}_1, \hat{k}_2, \hat{\rho}_b)$ is given in Appendix 5.2. Since all resulting parameter estimates are continuous functions of the cell frequencies, it follows that $\sqrt{N}(\hat{\omega} - \omega)$ has an approximate multivariate normal distribution with mean vector 0 and variance covariance matrix V.

Example 5.7

Over the past 30 years, there has been an increasing emphasis on the development of screening instruments to identify mental disorders in primary care settings, including anxiety and depression.

However, it has been suggested by psychiatric researchers that primary care patients tend to formulate their problems in somatic terms, not only to physicians but also to themselves, making it difficult to detect the underlying disorder.

The data in this example presented in a study reported by Becker et al. (2002), who used two instruments; the Patient Health Questionnaire (PHQ) and the Structured Clinical Interview (SCI) to document their utility as tools for diagnosing anxiety and depression. The first 173 patients attending a teaching hospital were interviewed by an experienced clinical psychiatrist immediately following completion of the PHQ and the scheduled primary care physician visit. The resulting data are summarized in Table 5.24.

TABLE 5.24

Cross Tabulation of (a) Anxiety and (b) Depression, by PHQ and SCI and (c) Number of Patients on whom 0, 1, or 2 Traits, Are Rated as Present. (0 = absent, 1 = present)

SCI	0		1	Total
(a) PHQ				
0	113		19	132
1	3		38	41
Total	116		57	173
(b)				
0	118		15	133
1	6		34	40
Total	124		49	173
SCI X_1	0	1	2	Total
(c) PHQ X_2				
0	93	10	10	113
1	11	6	5	22
2	14	5	19	38
Total	118	21	34	173

The estimates obtained from these data are given by $\hat{\pi}_1 - 0.283$, $\hat{k}_1 = 0.6900$, $\hat{\pi}_2 = 0.257$, $\hat{k}_2 = 0.6823$, and $\hat{\rho}_b = 0.3718$, with the elements of the estimated variance–covariance matrix calculated using the expressions in Appendix 5.3 as

$$\widehat{\text{var}}\left(\hat{\pi}_1\right) = 0.001, \widehat{\text{var}}\left(\hat{\pi}_2\right) = 0.001, \widehat{\text{var}}\left(\hat{k}_1\right) = 0.0038,$$

$$\widehat{\text{var}}\left(\hat{k}_2\right) = 0.004, \widehat{\text{var}}\left(\hat{\rho}_b\right) = 0.0043,$$

$$\widehat{\text{cov}}\left(\hat{\pi}_1,\hat{\pi}_2\right) = 0.0004, \widehat{\text{cov}}\left(\hat{\pi}_1,\hat{k}_1\right) = 0.0003, \widehat{\text{cov}}\left(\hat{\pi}_1,\hat{k}_2\right) = 0.0002,$$

$$\widehat{\text{cov}}\left(\hat{\pi}_1,\hat{\rho}_b\right) = 0.0003, \widehat{\text{cov}}\left(\hat{\pi}_2,\hat{k}_1\right) = 0.0001, \widehat{\text{cov}}\left(\hat{\pi}_2,\hat{k}_2\right) = 0.0004,$$

$$\widehat{\text{cov}}\left(\hat{\pi}_2,\hat{\rho}_b\right) = 0.0004, \widehat{\text{cov}}\left(\hat{k}_1,\hat{k}_2\right) = 0.0007, \widehat{\text{cov}}\left(\hat{k}_1,\hat{\rho}_b\right) = 0.006,$$

and

$$\widehat{\text{cov}}\left(\hat{k}_2,\hat{\rho}_b\right) = 0.0066.$$

Approximately 95% confidence intervals about k_1 and k_1 may be obtained from the estimated standard errors of $\hat{k}_1$ and $\hat{k}_2$. These are given by (0.569, 0.811) and (0.558, 0.806), respectively. Furthermore, a 95% confidence interval about $\hat{k}_1 - \hat{k}_2$ may be obtained as $0.0077 \pm 1.96 \{0.0038 + 0.004 - 2(0.0007)\}^{1/2}$ or (−0.149, 0.165), indicating that the two estimated coefficients are not significantly different.

The overall level of agreement between the two instruments may therefore be obtained by constructing a pooled estimator. We consider two such estimators: the first proposed by Donner et al. (2000), and the other which assigns weights that minimize the variance of the pooled estimator.

These estimators are given, respectively, by

$$\bar{k}_\lambda = \lambda k_1 + (1 - \lambda)k_2$$

and

$$\bar{k}_\omega = \omega \bar{k}_1 + (1 - \omega)\bar{k}_2,$$

where

$$\lambda = \frac{\hat{\pi}_1\left(1 - \hat{\pi}_1\right)}{\hat{\pi}_1\left(1 - \hat{\pi}_1\right) + \hat{\pi}_2\left(1 - \hat{\pi}_2\right)}.$$

It can be easily shown that the value of ω that minimizes $\text{var}(\hat{k}_\omega)$ is given by

$$\omega = \frac{\upsilon_2 - \upsilon_{12}}{\upsilon_1 + \upsilon_2 - 2\upsilon_{12}},$$

where $\upsilon_j = \text{var}(\hat{k}_j)$ and $\upsilon_{12} = \text{cov}(\hat{k}_1, \hat{k}_2)$. The minimized value of $\text{var}(\hat{k}_\omega)$ is then given by

$$\bar{k}_\lambda = 0.68463, \text{ with standard error } SE(\hat{k}_\lambda) = 0.0479$$

and

$$\bar{k}_\omega = 0.68467, \text{ with standard error } SE(\bar{k}_\omega) = 0.0405.$$

It is therefore, seen that the relatively simple estimator $\bar{k}_\lambda$ yields virtually the same results as the minimum variance estimator $\bar{k}_\omega$. We also note that the value 0.68 may be characterized as "substantial," using guidelines proposed by Landis and Koch (1977).

The model can be extended to the situation where multiple traits are assessed by randomly selected two raters. In the following section, we describe the extension of the model.

5.11.2 Simultaneous Modeling of Interrater and Intrarater Agreements

The above model is flexible enough to model both interrater and intrarater agreements. To measure interrater agreement each rater must evaluate each subject in the sample at least twice. The samples need not be the same from one rater to another. However, to evaluate interrater agreement, each rater must assess the same sample of subjects. To demonstrate how the above model can be sued to assess both measures we consider data presented by Powell et al. (1999) analyzed by Kirchner and Lemke (2002). The data are from a mammography study comparing the equivalence between film-screen and digital images. As part of the study, five radiologists scored five regions of each breast on a five-point scale (normal, benign, probably benign, suspicious, and malignant). A repeat assessment was done on the film-screen mammograms under blinded and identical experimental conditions. Following Kirchner and Lemke (2002), normal, benign, and probably benign were considered negative (category = 0) while suspicious and malignant were considered positive (category = 1). The raw data are presented in Table 5.25.

Two sets of estimates of agreement may be obtained. The first are the intraclass kappa statistics, and these are given in Tables 5.25a and 5.25b.

The intrarater reliability of rater B is somewhat higher than that of rater A.

The other set consists of pairwise kappa statistics measuring the interrater agreements (see Tables 5.25c_11, 5.25c_12, 5.25c_21, and 5.25c_22).

TABLE 5.25

Mammography Data

A1	A2	B1	B2	count
1	1	1	1	6
1	1	1	0	0
1	1	0	1	2
1	1	0	0	1
1	0	1	1	1
1	0	1	0	0
1	0	0	1	0
1	0	0	0	3
0	1	1	1	1
0	1	1	0	0
0	1	0	1	0
0	1	0	0	2
0	0	1	1	1
0	0	1	0	3
0	0	0	1	0
0	0	0	0	38

Note: A1 are the ratings by rater A on the first occasion A2 are the ratings by rater A on the second occasion. The same notation is used for rater B.

TABLE 5.25a

Cross Classifications of Rater A for the Two Occasions

A1	A2 0	1	Total
0	42	3	45
1	4	9	13
Total	46	12	58

Note: Kappa = 0.6432 (0.1233).

TABLE 5.25b

Cross Classifications of Rater B for the Two Occasions

B1	B2 0	1	Total
0	44	2	46
1	3	9	12
Total	47	11	58

Note: Kappa = 0.7290 (0.1141).

TABLE 5.25c_1.1

Cross Classifications of Raters A and B by
Occasion

	B1		
A1	0	1	Total
0	40	5	45
1	6	7	13
Total	46	12	58

Note: Kappa = 0.4394 (0.1422).

TABLE 5.25c_1.2

Cross Classifications of Raters A and B by
Occasion

	B2		
A1	0	1	Total
0	43	2	45
1	4	9	13
Total	47	11	58

Note: Kappa = 0.6854 (0.1189).

TABLE 5.25c_2.1

Cross Classifications of Raters A and B by
Occasion

	B1		
A2	0	1	Total
0	41	5	46
1	5	7	12
Total	46	12	58

Note: Kappa = 0.4746 (0.1422).

TABLE 5.25c_2.2

Cross Classifications of Raters A and B by
Occasion

	B2		
A2	0	1	Total
0	44	2	46
1	3	9	12
Total	47	11	58

Note: Kappa = 0.7290 (0.1141).

TABLE 5.26

Cross Classification of Total Scores

	y = (B1 + B2)			
x = (A1 + A2)	0	1	2	Total
0	38	3	1	42
1	5	0	2	7
2	1	2	6	9
Total	44	5	9	58

Note: Simple Kappa = 0.4200 (0.1089)
Weighted Kappa = 0.5731 (0.1050).

There seems to be a higher agreement between A1–B2, and A2–B2. One can also measure the overall agreement between the two raters averaged over the replications from the following 3×3 table (Table 5.26).

The model based gives an overall estimate of agreement very close to the weighted Cohen's kappa (κ_b) = 0.587 (0.101). Since the agreement statistic for rater A, $\kappa_A = 0.643$ (0.123) and that of rater B, $\kappa_B = 0.729$ (0.114), and cov(κ_A, κ_B) = −0.001229, we can test the equality of the two agreement coefficients using the large sample properties of their estimates. Formally, we would like to test $H_0 : \kappa_A = \kappa_B$, which can be tested by referring the Z-score:

$$Z = \frac{\hat{\kappa}_A - \hat{\kappa}_B}{\sqrt{SE^2(\hat{\kappa}_A) + SE^2(\hat{\kappa}_B) - 2\text{cov}(\hat{\kappa}_A, \hat{\kappa}_B)}} = \frac{0.729 - 0.643}{\sqrt{0.013 + 0.015 + 0.001229}} = 0.491.$$

The p-value associated with the test statistic is 0.64. Indicating that, overall, the two raters have the same level of agreement.

EXERCISES

E5.1 Consider the data in Chapter 2, which can be found in the attached CD. The data are from an investigation of agreement between two raters with respect to the angiographic classification of lumen narrowing in the internal carotid artery. The degree of narrowing in each of 125 patients may be classified into severe narrowing (greater than or equal to 70%), and nonsevere narrowing (less than 70%). Here are the data lines of the first five subjects:

```
input x11 x12 x21 x22;
cards;
   23      25      25      26
   52      71      70      58
   49      40      59      51
   16      21      23      27
   65      65      70      66
```

You need to recode the data such that
$y_{ij} = 1$ if $x_{ij} \leq 70$, and $y_{ij} = 0$ otherwise. Define $z_1 = y_{11} + y_{12}$, and
$z_2 = y_{21} + y_{22}$. Use SAS to produce a cross classification for z_1 by z_2.

E5.2 Find the MLE of the prevalence parameter and the two coefficients intraclass kappa Equations 5.27a and 5.27.b. Moreover, use equation $\hat{\kappa}_b = ((n_{11} + 2n_{12} + 2n_{21} + 4n_{22}) - 4N\hat{\pi}^2/4N\hat{\pi}(1 - \hat{\pi}))$ to estimate the cross correlation parameter.

E5.3 Use the results in Appendix 5.I to derive the asymptotic covariance matrix of the vector of estimates $\hat{M} = (\hat{\pi}, \hat{\rho}_1, \hat{\rho}_2)$.

E5.4 For continuous measurements we have shown that the asymptotic variance of the ICC is given by $\text{var}(\hat{\rho}) = (2(1 - \rho)^2(1 + (n - 1)\rho)^2/kn(n - 1))$. After dichotomization, the variance of the test–retest index of reliability becomes

$$\text{var}(\hat{\kappa}) = \frac{1 - \kappa}{N}\left[(1 - \kappa)(1 - 2\kappa) + \frac{\kappa(2 - \kappa)}{2\pi(1 - \pi)}\right].$$

Define the cost of dichotomization by the ratio of the above two variances. In other words, it is the relative efficiency of $\hat{\kappa}$ relative to $\hat{\rho}$. Comment on the situations when the cost of such dichotomization is high. Take $\rho = \kappa$ and the suggested values of $\pi = 0.1, 0.2, 0.3, 0.4, 0.5$. Note that $n = 2$ and $N = nk$.

E5.5 In a typical 2×2 classification, define the estimated PABAK as

$$PABAK = 2(n_{11} + n_{22})/n - 1.$$

Find the exact variance of the above estimator using the well known moments of n_{ij} using the properties of the multinomial distribution.

E5.6 Hirji and Rosove (1990) derived the large sample variance and covariance of $(\hat{\lambda}_1, \hat{\lambda}_2)$ as $n^{-1}v(\hat{\lambda}_i)$, and $n^{-1}c(\hat{\lambda}_1, \hat{\lambda}_2)$, where

$$v(\hat{\lambda}_i) = p_{ii}\{(p_{i.} - p_{ii})p_{i.}^{-3} + 2(p_{i.} - p_{ii})(p_{.i} - p_{ii})p_{i.}^{-2}p_{.i}^{-2} + (p_{.i} - p_{ii})p_{.i}^{-3}\}.$$

For $i \neq j$, they gave

$$c(\hat{\lambda}_i, \hat{\lambda}_j) = p_{ii}p_{jj}(p_{ij}p_{i.}^{-2}p_{.j}^{-2} + p_{ji}p_{.i}^{-2}p_{.j}^{-2}).$$

Use the above results to find the asymptotic variance of the estimator $\hat{\lambda} = (\hat{\lambda}_1 + \hat{\lambda}_2)/2$ and construct 95% confidence interval on the respective parameter. Note that the estimator of the variances and covariance is obtained when p_{ij} is replaced with n_{ij}/n, $p_{i.}$ is replaced with $n_{i.}/n$, and $p_{.j}$ is replaced with $n_{.j}/n$.

Appendix 5.1

Joint probability distribution of X_{i11}, X_{i12}, X_{i21}, X_{i22}

$$P(0,0,0,0) = \Delta^{-1}\left[(b+1)(b+2) + (\kappa_{c1} + \kappa_{c2})ab(b+1)(b+2)\right.$$
$$\left. + \kappa_{c1}\kappa_{c2}ab(a+1)(b+1)\right] = \Theta_{11},$$

$$P(1,0,0,0) = P(0,1,0,0) = 2\Delta^{-1}(1-k_{c1})\left[ab(b+1)(b+2)\right.$$
$$\left. + k_{c2}ab(a+1)(b+1)\right] = \Theta_{12},$$

$$P(1,1,0,0,) = \Delta^{-1}\left[(1+k_{c1}k_{c2})ab(a+1)(b+1) + k_{c1}ab(b+1)(b+2)\right.$$
$$\left. + k_{c2}ab(a+1)(a+2)\right] = \Theta_{13},$$

$$P(0,0,1,0) = P(0,0,0,1) = 2\Delta a^{-1}(1-k_{c2})\left[ab(b+1)(b+2)\right.$$
$$\left. + k_{c1}ab(a+1)(b+1)\right] = \Theta_{21},$$

$$P(1,0,1,0) = P(0,1,1,0) = 4\Delta^{-1}(1-k_{c1})(1-k_{c2})\left[ab(a+1)(b+1)\right] = \Theta_{22},$$

$$P(0,1,1,1) = P(1,0,1,1) = 2\Delta^{-1}(1-k_{c2})\left[ab(a+1)(a+2)\right.$$
$$\left. + k_{c1}ab(a+1)(b+1)\right] = \Theta_{23},$$

$$P(0,0,1,1) = \Delta^{-1}\left[(1+k_{c1}k_{c2})ab(a+1)(b+1) + k_{c1}ab(a+1)(a+2)\right.$$
$$\left. + k_{c2}ab(b+1)(b+2)\right] = \Theta_{31},$$

$$P(1,1,1,0) = P(1,1,0,1) = 2\Delta_{-1}(1-k_{c1})\left[ab(a+1)(a+2)\right.$$
$$\left. + k_{c2}ab(a+1)(b+1)\right] = \Theta_{32},$$

$$P(1,1,1,1) = \Delta^{-1}\left[a(a+1)(a+2)(a+3) + (k_{c1}+k_{c2})ab(a+1)(a+2)\right.$$
$$\left. + k_{c1}k_{c2}ab(a+1)(b+1)\right] = \Theta_{33},$$

$$P_1(k_1) = \frac{b(b+1)}{(a+b)(a+b+1)} + k_{c1}\frac{ab}{(a+b)(a+b+1)},$$

$$P_2(k_1) = \frac{2(1-k_{c1})ab}{(a+b)(a+b+1)},$$

$$P_3(k_1) = \frac{a(a+1)}{(a+b)(a+b+1)} + k_{c1}\frac{ab}{(a+b)(a+b+1)},$$

$$P_1(k_2) = \frac{b(b+1)}{(a+b)(a+b+1)} + k_{c2}\frac{ab}{(a+b)(a+b+1)},$$

$$P_2(k_2) = \frac{2(1-k_{c2})ab}{(a+b)(a+b+1)},$$

$$P_3(k_2) = \frac{a(a+1)}{(a+b)(a+b+1)} + k_{c2}\frac{ab}{(a+b)(a+b+1)},$$

where

$$\Delta = (a+b)(a+b+1)(a+b+2)(a+b+3),$$

$$a = (\pi(1-kb)/kb), b = (\pi(1-\pi)(1-k_b)/k_b),$$

and

$$k_b = (1+a+b)-1.$$

Correlation between $\hat{\kappa}_1$ and $\hat{\kappa}_2$.

Since the $\hat{\kappa}_j$ are functions of $M = (V_{11}, V_{12}, V_{13}, V_{21}, V_{22}, V_{23}, V_{31}, V_{32}, V_{33})'$, then to the first order of approximation, application of the delta method gives

$$\mathrm{cov}\left(\hat{\kappa}_1, \hat{\kappa}_2\right) = \sum_{ijlm} \mathrm{cov}\left(V_{ij}, V_{lm}\right)\left(\frac{\partial\hat{\kappa}_1}{\partial n_{ij}}\right)\left(\frac{\partial\hat{\kappa}_2}{\partial n_{lm}}\right).$$

Since M has a multinomial distribution, then $\mathrm{cov}(V_{ij}, V_{lm}) = -N\theta_{ij}\theta_{lm}$ $i \neq l$, $j \neq m$, $\mathrm{var}(V_{ij}) = N\theta_{ij}(1-\theta_{ij})$ and $\mathrm{cov}(n_{i1}, n_{j2}) = N[\theta_{il} - P_i(\kappa_1)P_j(\kappa_2)]$ $i, j = 1, 2, 3$. Hence, after some algebra, we obtain

$$N\,\mathrm{cov}(\hat{\kappa}_1, \hat{\kappa}_1) = d_1A - d_2\left(\frac{A}{2}+B\right) - d_3\left(\frac{A}{2}+C\right) + d_4\left(A + 2B + 2C + 4D\right),$$

where

$$d_1 = \left[4\pi^2(1-\pi)^2\right]^{-1}, \quad d_2 = \frac{P_2(\kappa_1)(1-2\pi)}{4\pi^3(1-\pi)^3},$$

$$d_3 = \frac{P_2(\kappa_2)(1-2\pi)}{4\pi^3(1-\pi)^3}, \quad d_4 = \frac{P_2(\kappa_1)P_2(\kappa_2)(1-2\pi)}{16\pi^4(1-\pi)^4},$$

$$A = \Theta_{22} - P_2(\kappa_1) P_2(\kappa_2),$$

$$B = \Theta_{32} - P_3(\kappa_2) P_2(\kappa_1),$$

$$C = \Theta_{23} - P_3(\kappa_1) P_2(\kappa_2),$$

$$D = \Theta_{33} - P_3(\kappa_2) P_3(\kappa_1).$$

A sample estimate of

$$\widehat{\text{corr}}(\hat{\kappa}_1, \hat{\kappa}_2) = \frac{\widehat{\text{cov}}(\hat{\kappa}_1, \hat{\kappa}_2)}{\sqrt{\left\{\widehat{\text{var}}(\hat{\kappa}_1) \widehat{\text{var}}(\hat{\kappa}_2)\right\}}},$$

where $\widehat{\text{cov}}(\hat{\kappa}_1, \hat{\kappa}_2)$ is obtained by replacing κ_j by $\hat{\kappa}_j$, κ_b by $\hat{\kappa}_b$ and π by $\hat{\pi}$ in terms of right hand side of $N\widehat{\text{cov}}(\hat{\kappa}_1, \hat{\kappa}_2)$.

Appendix 5.2

Let (P_1, P_2) be a random variable with specified marginals such that P_j has a β-distribution with the probability density function

$$f(p_j) \equiv \beta \left(\frac{(1 - \rho_b)\pi_j}{\rho_b}, \frac{(1 - \rho_b)(1 - \pi_j)}{\rho_b} \right).$$

The bivariate distribution for (P_1, P_2) using the canonical representation has density function $g(p_1, p_2)$ we can express

$$f(p_1, p_2) = f(p_1) f(p_2) \left[1 + \rho_b \frac{(p_1 - E(p_1))(p_2 - E(p_2))}{\sqrt{\text{var}(p_1) \text{var}(p_2)}} \right], \quad 0 \le p_j \le 1$$

From the definition of the noncentral product moments

$$\psi(r,s) = E\left(P_1^r P_2^s \right) = \int_0^1 \int_0^1 p_1^r p_2^s g(p_1, p_2) \, dp_1 dp_2.$$

We can easily show that

$$\vartheta_{22} = \psi(2,2)(1-\rho_1)(1-\rho_2) + \psi(1,2)\rho_1(1-\rho_2)$$
$$+\psi(2,1)\rho_2(1-\rho_1) + \psi(1,1)\rho_1\rho_2,$$

$$\vartheta_{12} = 2\{\psi(1,1)\rho_1(1-\rho_2) + \psi(2,1)(1-\rho_1)(1-\rho_2) - \psi(1,2)\rho_1(1-\rho_2)$$
$$- \psi(2,2)(1-\rho_1)(1-\rho_2)\},$$

$$\vartheta_{02} = \psi(2,0)(1-\rho_1) - \psi(2,1)(1-\rho_1)(1-\rho_2) + \psi(1,2)\rho_1(1-\rho_2)$$
$$-\psi(1,1)\rho_1(2-\rho_2) + \psi(2,2)(1-\rho_1)(1-\rho_2) + \psi(1,0)\rho_1,$$

$$\vartheta_{21} = 2\{\psi(1,1)\rho_2(1-\rho_2) + \psi(1,2)(1-\rho_1)(1-\rho_2) - \psi(2,1)\rho_2(1-\rho_1)$$
$$-\psi(2,2)(1-\rho_1)(1-\rho_2)\},$$

$$\vartheta_{11} = 4\{\psi(1,1)(1-\rho_1)(1-\rho_2) - \psi(2,1)(1-\rho_1)(1-\rho_2)$$
$$- \psi(1,2)(1-\rho_1)(1-\rho_2)+\psi(2,2)(1-\rho_1)(1-\rho_2)\},$$

$$\vartheta_{01} = 2\{\psi(1,0)(1-\rho_1) - \psi(1,1)(1-\rho_1)(2-\rho_2) + \psi(1,2)(1-\rho_1)(1-\rho_2)$$
$$-\psi(2,0)(1-\rho_1) + \psi(2,1)(1-\rho_1)(2-\rho_2) - \psi(2,2)(1-\rho_1)(1-\rho_2)\},$$

$$\vartheta_{20} = \psi(0,1)\rho_2 + \psi(0,2)(1-\rho_2) - \psi(1,2)(1-\rho_2)(2-\rho_1)$$
$$+ \psi(2,2)(1-\rho_1)(1-\rho_2) - \psi(1,1)\rho_2(2-\rho_1) + \psi(2,1)\rho_2(1-\rho_1),$$

$$\vartheta_{10} = 2\{\psi(0,1)(1-\rho_2) - \psi(0,2)(1-\rho_2) - \psi(1,1)(2-\rho_1)(1-\rho_2)$$
$$+ \psi(1,2)(2-\rho_1)(1-\rho_2) + \psi(2,1)(1-\rho_1)(2-\rho_2)$$
$$-\psi(2,2)(1-\rho_1)(1-\rho_2)\},$$

$$\vartheta_{00} = 1 - \psi(1,0)(2-\rho_1) - \psi(0,1)(2-\rho_2)$$
$$+ \psi(2,0)(1-\rho_1) + \psi(0,2)(1-\rho_2) + \psi(1,1)(2-\rho_1)(2-\rho_2)$$
$$- \psi(2,1)(1-\rho_1)(2-\rho_2) - \psi(1,2)(1-\rho_2)(2-\rho_1)$$
$$+ \psi(2,2)(1-\rho_1)(1-\rho_2).$$

Appendix 5.3

Elements of the asymptotic variance–covariance matrix of the estimated parameters are derived using the formulae for the moments of the sample moments, followed by several applications of the delta method. Following Kendall and Ord (1991), we write

$$\operatorname{cov}\left(m'_{r,s}, m'_{u,v}\right) = E\left(X_1^{r+u} X_2^{s+v}\right) - E\left(X_1^{r+s}\right) E\left(X_2^{u+v}\right).$$

Define the noncentral bivariate moments by $\mu'_{rs} = E[(X_1)^r (X_2)^s]$ and their sample moments estimator by $m'_{rs} = (1/N) \sum_1^N X_{i1}^r X_{i2}^s$.
Covariance between π_j and k_l for $j \neq l$ is

$$\operatorname{cov}\left(\hat{\pi}_1, \hat{k}_2\right) = \left(4\pi_2\left(1 - \pi_2\right)\right)^{-1}\left[-\operatorname{cov}\left(m'_{10}, m'_{01}\right)\left(4\pi_2 + \left(1 - 2\pi_2\right)\left(1 - k_2\right)\right)\right.$$
$$\left. + \operatorname{cov}\left(m'_{10}, m'_{02}\right)\right].$$

By symmetry,

$$\operatorname{cov}\left(\hat{\pi}_2, \hat{k}_1\right) = \left(4\pi_1\left(1 - \pi_1\right)\right)^{-1}\left[-\operatorname{cov}\left(m'_{10}, m'_{01}\right)\left(4\pi_1 + \left(1 - 2\pi_1\right)\left(1 + k_1\right)\right)\right.$$
$$\left. + \operatorname{cov}\left(m'_{01}, m'_{20}\right)\right].$$

Let

$$a_j = -\frac{4\pi_j + \left(1 - 2\pi_j\right)\left(1 + k_j\right)}{2\pi_j\left(1 - \pi_j\right)} \quad \text{and} \quad b_j = \frac{1}{2\pi_j\left(1 - \pi_j\right)} \, j = 1, \, 2.$$

By the delta method we show to the first order of approximation that

$$\operatorname{cov}\left(\hat{k}_1, \hat{k}_2\right) = a_1 a_2 \operatorname{cov}\left(m'_{10}, m'_{01}\right) + a_1 b_2 \operatorname{cov}\left(m'_{10}, m'_{02}\right) + a_2 b_1 \operatorname{cov}\left(m'_{01}, m'_{20}\right)$$
$$+ a_2 b_2 \operatorname{cov}\left(m'_{01}, m'_{02}\right) + b_1 b_2 \operatorname{cov}\left(m'_{20}, m'_{02}\right),$$

$$\mathrm{var}(\hat{\rho}_b) = \mathrm{var}(m'_{11})(D\rho_{b11})^2 + \mathrm{var}(m'_{10})(D\rho_{b10})^2 + \mathrm{var}(m'_{01})(D\rho_{b01})^2$$
$$+ 2\big[\mathrm{cov}(m'_{10}, m'_{01})(D\rho_{b10})(D\rho_{b01}) + \mathrm{cov}(m'_{10}, m'_{11})(D\rho_{b10})(D\rho_{b01})$$
$$+ \mathrm{cov}(m'_{01}, m'_{11})(D\rho_{b01})(D\rho_{b10})\big];$$

here

$$D\rho_{b11} = \frac{1}{4\tau},$$

$$D\rho_{b10} = \rho_b(2\pi_1 - 1)/4\pi_1(1 - \pi_1) - \pi_2/2\tau,$$

and

$$D\rho_{b01} = \rho_b(2\pi_2 - 1)/4\pi_2(1 - \pi_2) - \pi_1/2\tau,$$

$$\mathrm{cov}\left(\hat{k}_1, \hat{\rho}_b\right) = \mathrm{var}(m'_{10})(D\rho_{b10})a_1 + \mathrm{cov}(m'_{10}, m'_{11})(D\rho_{b11})a_1$$
$$+ \mathrm{cov}(m'_{20}, m'_{10})(D\rho_{b10})b_1 + \mathrm{cov}(m'_{20}, m'_{11})(D\rho_{b11})b_1,$$

$$\mathrm{cov}\left(\hat{k}_2, \hat{\rho}_b\right) = \mathrm{var}(m'_{01})(D\rho_{b01})a_2 + \mathrm{cov}(m'_{01}, m'_{11})(D\rho_{b11})a_2$$
$$+ \mathrm{cov}(m'_{02}, m'_{01})(D\rho_{b01})b_2 + \mathrm{cov}(m'_{02}, m'_{11})(D\rho_{b11})b_2,$$

$$\mathrm{cov}\left(\hat{\pi}_1, \hat{\rho}_b\right) =$$

$$\frac{1}{2}\big[\mathrm{var}(m'_{10})(D\rho_{b10}) + \mathrm{cov}(m'_{10}, m'_{01})(D\rho_{b01}) + \mathrm{cov}(m'_{10}, m'_{11})(D\rho_{b11})\big],$$

$$\mathrm{cov}\left(\hat{\pi}_2, \hat{\rho}_b\right) =$$

$$\frac{1}{2}\big[\mathrm{var}(m'_{01})(D\rho_{b01}) + \mathrm{cov}(m'_{10}, m'_{01})(D\rho_{b10}) + \mathrm{cov}(m'_{01}, m'_{11})(D\rho_{b11})\big],$$

$$\mathrm{corr}\left(\hat{k}_1, \hat{k}_2\right) = \mathrm{cov}\left(\hat{k}_1, \hat{k}_2\right)\Big/\sqrt{\mathrm{var}\left(\hat{k}_1\right)\mathrm{var}\left(\hat{k}_2\right)},$$

$$\text{var}\left(\overline{k}_\lambda\right) = \lambda^2 \upsilon_1 + (1-\lambda)^2 \upsilon_2 + 2\lambda(1-\lambda)\upsilon_{12}$$

$$+ \left(k_1 - k_2\right)^2 \frac{\lambda^2 \left(1-\lambda\right)^2 \left(1-2\pi_1\right)^2}{\pi_1^2 \left(1-\pi_1\right)^2} \text{var}\left(\hat{\pi}_1\right)$$

$$+ \left(k_1 - k_2\right)^2 \frac{\lambda^4 \left(1-2\pi_2\right)^2}{\pi_1^2 \left(1-\pi_1\right)^2} \text{var}\left(\hat{\pi}_2\right)$$

$$+ 2\left(k_1 - k_2\right) \frac{\lambda^2 \left(1-\lambda\right)\left(1-2\pi_1\right)}{\pi_1 \left(1-\pi_1\right)} \text{cov}\left(\hat{\pi}_1, \hat{k}_1\right)$$

$$- 2\left(k_1 - k_2\right) \frac{\lambda^3 \left(1-2\pi_2\right)}{\pi_1 \left(1-\pi\right)} \text{cov}\left(\hat{\pi}_2, k_1\right)$$

$$+ 2\left(k_1 - k_2\right) \frac{\lambda \left(1-\lambda\right)^2 \left(1-2\pi_1\right)}{\pi_1 \left(1-\pi_1\right)} \text{cov}\left(\hat{\pi}_1, k_2\right)$$

$$- 2\left(k_1 - k_2\right) \frac{\lambda^2 \left(1-\lambda\right)\left(1-2\pi_2\right)}{\pi_1 \left(1-\pi\right)} \text{cov}\left(\hat{\pi}_2, \hat{k}_2\right)$$

$$- 2\left(k_1 - k_2\right)^2 \frac{\lambda^3 \left(1-\lambda\right)\left(1-2\pi_1\right)\left(1-2\pi_2\right)}{\pi_1^2 \left(1-\pi_1\right)^2} \text{cov}\left(\hat{\pi}_1, \pi_2\right).$$

6

Coefficients of Agreement for Multiple Rates and Multiple Categories

6.1 Introduction

In Chapter 5, we investigated the issue of agreement between the two raters, when each classifies a set of n subjects into one of two nominal scale categories (e.g., yes/no; disease/no-disease, etc.). In this chapter, we extend the discussion on the subject of agreement to situations when we have

1. Multiple categories and two raters
2. Multiple raters and two categories
3. Multiple raters and multiple categories

Much of the work in this chapter is based on the seminal work of Agresti (1992), Cohen (1968), Kraemer (1979), and many others.

6.2 Multiple Categories and Two Raters

6.2.1 Category Distinguishability

Let n subjects be classified into c nominal scale Categories 1, 2, ..., c by two raters using a single rating protocol, and let π_{jk} be the joint probability that the first rater classifies a subject as j and the second rater classifies the same subject as k. Let $\pi_{j.} = \sum_k \pi_{jk}$ and $\pi_{.k} = \sum_j \pi_{jk}$. This above set-up is depicted in the $c \times c$ classification (Table 6.1). There are two questions; the simpler one concerns the interrater bias, or the difference between the two sets of marginal probabilities $\pi_{j.}$ and $\pi_{.j}$, while the second is related to the magnitude of $\sum_j \pi_{jj}$, or the extent of agreement of the two raters about individual subjects.

The marginal probabilities $\pi_{i.} = \sum_{j=1}^{c} \pi_{ij}$ and $\pi_{.j} = \sum_{i=1}^{c} \pi_{ij}$ describe how the raters independently allocate subjects to the assigned categories. Discrepancies among the respective marginal probabilities (i.e., whenever $\pi_{i.} \neq \pi_{.i}$) indicate

TABLE 6.1

Joint and Marginal Probabilities of Classification by Two Raters

	Rater (1)				
Rater (2)	1	2	...	c	Total
1	π_{11}	π_{12}		π_{1c}	$\pi_{1\cdot}$
2	π_{21}	π_{22}		π_{2c}	$\pi_{2\cdot}$
$\vdots$	$\vdots$				
c	$\pi_{c}1$	$\pi_{c}2$		π_{cc}	$\pi_{c\cdot}$
Total	$\pi_{\cdot1}$	$\pi_{\cdot2}$		$\pi_{\cdot c}$	1

the presence of relative bias. This means that bias is equivalent to marginal heterogeneity.

Cohen (1960) proposed the coefficient of agreement κ defined by

$$\kappa = \frac{\sum_{j=1}^{c} (\pi_{jj} - \pi_{j\cdot}\pi_{\cdot j})}{1 - \sum_{j=1}^{c} \pi_{j\cdot}\pi_{\cdot j}} \tag{6.1}$$

as a measure of agreement between two raters. Cohen had two justifications: first, the sum of the diagonal probabilities $\pi_o = \sum_{j=1}^{c} \pi_{jj}$ is the probability that the two rates agree on the classification of a subject. Second, the probability that they "agree on chance" is $\sum_{j} \pi_{j\cdot}\pi_{\cdot j}$, and this probability should therefore be subtracted from the first. The division by $1 - \sum_{j} \pi_{j\cdot}\pi_{\cdot j}$ results in a coefficient, whose maximum value is one, is attained when the off diagonal elements in Table 6.1 are zeros. The estimated value of κ is obtained by substituting n_{jk}/n for π_{jk}, where n_{jk} is the observed frequency of the (j, k) cell.

The estimated variance of κ from $c \times c$ table was given by Fleiss et al. (1969) as

$$\hat{var}_F(\hat{\kappa}) = \frac{A + B - C}{n(1 - P_e)^2}, \tag{6.2}$$

where

$$A = \sum_{i=1}^{c} \hat{\pi}_{ii}[1 - (\hat{\pi}_{i\cdot} + \hat{\pi}_{\cdot i})(1 - \hat{k})]^2,$$

$$B = (1 - \hat{\kappa})^2 \sum_{i \neq j}\sum \hat{\pi}_{ij}(\hat{\pi}_{i\cdot} + \hat{\pi}_{\cdot j})^2,$$

$$C = \left[\hat{\kappa} - P_e(1 - \hat{\kappa})\right]^2,$$

and P_e is an estimate of the "chance agreement" obtained on replacing $\pi_{i.}$ and $\pi_{.j}$ by their sample estimators.

An approximate $100(1-\alpha)\%$ confidence interval for κ is $\hat{\kappa} \pm z_{\alpha/2}\sqrt{\hat{\text{var}}_F(\hat{\kappa})}$.

The definition of κ given in Equation 6.1 is suitable for $c \times c$ tables with "nominal" response categories. For "ordinal" response, Cohen (1968) introduced the weighted kappa, κ_w, to allow each (j, k) cell to be weighted according to the degree of agreement between the jth and kth categories. Assigning weights $0 \le w_{jk} \le 1$ to the (j, k) cell with $w_{jj} = 1$, Cohen's weighted kappa is

$$\hat{\kappa}_w = \frac{\sum_{j=1}^{c} \sum_{k=1}^{c} w_{jk}(\pi_{jk} - \pi_{j.}\pi_{.k})}{1 - \sum_{j=1}^{c} \sum_{k=1}^{c} w_{jk}\pi_{j.}\pi_{.k}} \tag{6.3}$$

and is interpretable as the proportion of weighted agreement corrected for chance. Note that the unweighted kappa is a special case of k_w with $w_{jj} = 1$ for $i = j$ and $w_{ij} = 0$ for $i \ne j$. If, on the other hand, the c categories form an ordinal scale, with the categories assigned the numerical values $1, 2, ..., c$, and if $w_{ij} = 1 - ((i - j)^2/(c - 1)^2)$, then $\hat{k}_w$ can be interpreted as an ICC for a two-way ANOVA computed under the assumption that the n subjects and the two raters are random samples from populations of subjects and raters (Fleiss and Cohen, 1973).

Fleiss et al. (1969) derived the formula for the asymptotic variance of $\hat{\kappa}_w$ as:

$$\text{var}_F(\hat{\kappa}_w) = \frac{1}{n(1 - P_e(w))^2}\left[\sum_{i=1}^{c}\sum_{j=1}^{c}\pi_{ij}w_{ij}^{*\,2} - \left[\hat{\kappa}_w - P_{e(w)}(1 - \hat{\kappa}_w)\right]^2\right],$$

where

$$w_{ij}^* = w_{ij} - (\bar{w}_{i.} + \bar{w}_{.j})(1 - \hat{\kappa}_w),$$

$$P_e(w) = \sum_{i=1}^{c}\sum_{j=1}^{c}w_{ij}.\hat{\pi}_{i.}.\hat{\pi}_{.j},$$

$$\bar{w}_{i.} = \sum_{j=1}^{c}w_{ij}.\hat{\pi}_{.j},$$

$$\bar{w}_{.j} = \sum_{i=1}^{c}w_{ij}.\hat{\pi}_{i.}.$$

Note that with the specific weights

$$w_{ij} = 1 - \frac{(1-j)^2}{(c-1)^2}$$

proposed by Fleiss and Cohen (1973), the estimate of weighted kappa, $\hat{\kappa}_w$ in Equation 6.3 reduces to

$$\hat{\kappa}_w = 1 - \frac{\sum_i \sum_j (1-j)^2 n_{ij}}{1/n \sum_i \sum_j n_{i.} n_{.j} (i-j)^2}.$$

Furthermore, since

$$(i-j)^2 = (i - \bar{x}_1)^2 + (j - \bar{x}_2)^2 + (\bar{x}_1 - \bar{x}_2)^2 + 2(\bar{x}_1 - \bar{x}_2)(i - \bar{x}_1)$$
$$- 2(\bar{x}_1 - \bar{x}_2)(j - \bar{x}_2) - 2(i - \bar{x}_1)(j - \bar{x}_2),$$

then

$$\sum_i \sum_j n_{ij} (i-j)^2 = ns_1^2 + ns_2^2 + n(\bar{x}_1 - \bar{x}_2)^2 - 2ns_{12}$$

and

$$\sum_i \sum_j n_{i.} n_{.j} (i-j)^2 = ns_1^2 + ns_2^2 + n(\bar{x}_1 - \bar{x}_2)^2,$$

where

$$n\bar{x}_1 = \sum_i in_{i.}, \quad n\bar{x}_2 = \sum_i jn_{.j}, \quad ns_1^2 = \sum_i n_{i.} (i - \bar{x}_1)^2, \quad ns_2^2 = \sum_j n_{.j} (j - \bar{x}_2)^2,$$

and

$$ns_{12} = \sum \sum n_{ij} (i - \bar{x}_2)(j - \bar{x}_2).$$

Substituting in $\hat{\kappa}_w$, we obtain

$$\hat{\kappa}_w = \frac{2rs_1 s_2}{s_1^2 + s_2^2 + (\bar{x}_2 - \bar{x}_2)^2}.$$

The above expression was given by Kreppendorff (1970). It should be noted that it is equivalent to the CCC of Lin (1989) used as a measure of agreement between two sets of continuous measurements as shown in Chapter 2.

It is then clear that when the marginal distributions of the two raters are identical, that is, when $\bar{x}_1 = \bar{x}_2$ and $s_1^2 = s_2^2$, then both agreement and association are equal. Therefore, high agreement requires that the marginal distributions should be close to each other and that the association parameter (Pearson's correlation coefficient in this case) should be high as well.

Darroch and McCloud (1986) noted that the interest in kappa and its variants lies not so much in describing how well two particular expert observers agree with each other as in measuring how well any expert observer can distinguish the categories from each other.

In the words of Davies and Fleiss (1982) what is at issue is "the precision of the classification process."

In many circumstances, the categories into which subjects are classified do not have precise objective definitions. Therefore, one must recognize first, that different expert raters may interpret the category definitions differently and, second, that categories will not be completely distinguishable from each other, even by the same observer. Darroch and McCloud (1986) examined the two features, rater difference and category distinguishability in full detail. Their model for the joint classification probabilities of a randomly selected subject by two raters incorporates the following features: the classification of a given subject by a given rater is allowed to be random; different raters can have different classification probabilities for the same subject; no multiplicative interaction is assumed to hold between the subject effects and the rater effects in the classification probabilities. Darroch and McCloud defined the degree of distinguishability between two categories from the joint classification probabilities for two raters. The degree of distinguishability of two categories does vary from one rater pair to another; however, the average degree of distinguishability δ appears to vary very little. They noted that the kappa index depends heavily on which pair of raters classifies a set of subjects, and for this reason, they recommend that

$$\delta = 1 - \frac{2}{c(c-1)} \sum_{j=1}^{c=1} \sum_{k=j+1}^{c} \frac{\pi_{jk}\pi_{kj}}{\pi_{jj}\pi_{kk}} \tag{6.4}$$

be used in place of kappa.

Note that (1) $\delta = 1$ if and only if all pairs of categories are completely distinguishable, then in that case, $\sum_{j=1}^{3} \pi_{jj} = 1$. (2) $\delta = 0$, if and only if all pairs of categories are completely indistinguishable which means that $\pi_{jk} = \pi_j \pi_k$ for all j, k. (3) δ lies between 0 and 1.

Kappa possesses (1) and not (2) and (3).

TABLE 6.2

Classification of 149 MS Patients by Two Raters

	Neurologist 2				
Neurologist 1	1	2	3	4	Total
1	38	5	0	1	44
2	33	11	3	0	47
3	10	14	5	6	35
4	3	7	3	10	23
Total	84	37	11	17	149

Note: From the data kappa = 0.21, SE = 0.0505, while $\delta = 0.746$.

Example 6.1

We examine the MS data reported by Westlund and Kurland (1953) and analyzed by Landis and Koch (1977). Two neurologists independently classified 149 patients into one of the following classes: 1 ≡ certain MS, 2 ≡ probable MS, 3 ≡ possible MS, and 4 ≡ doubtful, unlikely, or definitely not MS. The definitions of the four categories are only partially objective and the two neurologists interpreted them somewhat differently. This is evident by consideration of their marginal frequencies in Table 6.2.

Walter et al. (1988) reported on data from two studies that aimed at assessing the reliability of self-reported patient histories. The data came from two case-control studies, the first study of cervical carcinoma, and the second of cervical dysplasia. Cases and controls were both interviewed about their use of Pap smear in the previous 5 years. The details of the studies were reported in Clarke and Anderson (1979) and Clarke et al. (1985). In Tables 6.3a and 6.3b, we show the number of smears reported in the last 5 years.

The study by Walter et al. (1988) explained the poor agreement of Pap smear results because of differences in terminology. They suggested that accurate information could have been obtained if the physician's response rates were higher.

One can test the significance of the difference between two independent kappa coefficients by employing a large sample test on the hypothesis $H_0 : \kappa_1 = \kappa_2$. Under general conditions, we may assume that as the sample size gets larger, $\hat{\kappa}$ is asymptotically unbiased with variance that can be consistently estimated by Equation 6.2 and follows a normal distribution. Therefore,

TABLE 6.3a

Number of Smears Reported by Cases and Their Physicians

Physicians	0–1	2–3	4–5	6+
0–1	1	6	5	5
2–3	3	13	16	17
4–5	3	8	33	28
6+	2	2	11	29

Note: $n = 181$, $\kappa_w = 0.21$, SE = 0.0495.

TABLE 6.3b

Number of Smears Reported by Controls and Their Physicians

Physicians	0–1	2–3	4–5	6+
0–1	11	26	24	3
2–3	13	33	51	10
4–5	5	5	45	6
6+	1	4	2	2

Note: $n = 241$, $\kappa_w = 0.15$, SE $= 0.0399$.

$$Z = \left(\frac{\hat{\kappa}_1 - \hat{\kappa}_2}{\sqrt{\text{var}(\hat{\kappa}_1) + \text{var}(\hat{\kappa}_2)}} \right)$$

has asymptotically standard normal distribution, when the null hypothesis holds. Hence, for the data in Tables 6.3a and 6.3b we have

$$Z = \left(\frac{0.21 - 0.15}{\sqrt{SE^2 + SE^2}} \right) = \left(\frac{0.10}{0.0636} \right) = 1.57 \; (p = 0.13).$$

This shows that the two measures of interobserver agreement are not statistically significant from each other.

6.2.2 Test for Interrater Bias

Recall that in the 2×2 contingency table in Chapter 5, McNemar's (1947) statistics is used to test for interrater bias. If the sample size n is large, the McNemar statistic $X^2 = ((n_{10} - n_{01})^2 / n_{10} + n_{01})$ has approximately a chi-square distribution with one degree of freedom when the hypothesis of marginal homogeneity is true.

A number of authors have generalized McNemar's test to the comparison of the marginals of $c \times c$ contingency table.

Here, we use a test statistic known as Stuart–Maxwell (SM) statistic to test the hypothesis: $\pi_c = \pi_r$, where $\pi_r = (\pi_1, \pi_2, \ldots, \pi_c)$ and $\pi_c = (\pi_1, \pi_2, \ldots, \pi_c)$ are the marginal probabilities defined as:

$$\pi_{i.} = \sum_{j=1}^{c} \pi_{ij}, \quad \pi_{.j} = \sum_{j=1}^{c} \pi_{ij},$$

and all such parameters are defined in Table 6.1.

Let n_{ij} denote the number of observations in the ith row and jth column, and are defined in Table 6.4

$$n_{i.} = \sum_{j=1}^{c} n_{ij} \quad n_{.j} = \sum_{i=1}^{c} n_{ij}.$$

TABLE 6.4

Number of Observations in the ith Row and jth Column

Rater (2)	1	2	...	j	...	c	Total
1	n_{11}	n_{12}		n_{2j}		n_{1c}	$n_{1.}$
2	n_{21}	n_{22}		n_{1j}		n_{2c}	$n_{2.}$
i	n_{i1}	n_{i2}		n_{ij}		n_{ic}	$n_{i.}$
⋮							
c	n_{c1}	n_{c2}		n_{cj}		n_{cc}	$n_{c.}$
Total	$n_{.1}$	$n_{.2}$		$n_{.j}$		$n_{.c}$	n

Furthermore, define $d_i = n_{i.} - n_{.i}$ $(i = 1, 2, ..., c)$.

The statistics $\{d_i\}$ are crucial for determining whether the row marginal distribution π_r is equal to the column marginal distribution π_c. That is, d_i are used to test $\pi_{i.} = \pi_{.i}$ for each $(i = 1, 2, ..., c)$.

Under the null hypothesis $\pi_{i.} = \pi_{.i}$ $(i = 1, 2, ..., c)$, we have

$$E(d_i) = 0,$$

$$V_{ii} = \text{var}_{\circ}(d_i) = n(\pi_{i.} + \pi_{.i} - 2\pi_{ii}),$$

$$V_{ij} = \text{cov}_{\circ}(d_i, d_j) = -n(\pi_{ij} + \pi_{ji}).$$

The consistent estimators of V_{ii} and V_{ij} are given, respectively by

$$\hat{V}_{ii} = n_{i.} + n_{.i} - 2n_{ii},$$

$$\hat{V}_{ij} = -2(\bar{n}_{ij}),$$

where

$$\bar{n}_{ij} = \frac{(n_{ij} + n_{ji})}{2}.$$

Since $\sum_{i=1}^{c} d_i$, the covariance matrix $\{\hat{V}_{ij}\}$ is singular. Suppose that one of the $d_i s$, however, is dropped and that the covariance matrix of the remaining $c - 1$ $d_i s$ is formed. Let D represent the resulting vector of $d_i s$ and V the resulting covariance matrix. Then V is nonsingular, and if n is large enough, then the statistic

$$X_c^2 = D^T V^{-1} D \tag{6.5}$$

has, approximately a chi-square distribution with $c-1$ degrees of freedom. Large values of X_c^2 indicate that the data do not support the hypothesis of no interrater bias.

Fleiss and Everitt (1971) gave explicit expressions for Equation 6.5 when $c = 3$ and $c = 4$.

For $c = 3$, X_c^2 becomes

$$X_3^2 = \frac{\bar{n}_{23}d_1^2 + \bar{n}_{13}d_2^2 + \bar{n}_{12}d_3^2}{2\left(\bar{n}_{12}\bar{n}_{23} + \bar{n}_{12}\bar{n}_{13} + \bar{n}_{13}\bar{n}_{23}\right)},$$

and for $c = 4$,

$$X_4^2 = \frac{1}{2\Delta}\left[a_1 d_1^2 + a_2 d_2^2 + a_3 d_3^2 + a_4 d_4^2 + \bar{n}_{12}\bar{n}_{34}\left(d_1 + d_2\right)^2 \right.$$
$$\left. + \bar{n}_{13}\bar{n}_{24}\left(d_1 + d_3\right)^2 + \bar{n}_{14}\bar{n}_{23}\left(d_1 + d_4\right)^2\right],$$

where

$$a_1 = \bar{n}_{23}\bar{n}_{24} + \bar{n}_{23}\bar{n}_{34} + \bar{n}_{24}\bar{n}_{34},$$
$$a_2 = \bar{n}_{13}\bar{n}_{14} + \bar{n}_{13}\bar{n}_{34} + \bar{n}_{14}\bar{n}_{34},$$
$$a_3 = \bar{n}_{12}\bar{n}_{14} + \bar{n}_{12}\bar{n}_{24} + \bar{n}_{14}\bar{n}_{24},$$
$$a_4 = \bar{n}_{12}\bar{n}_{13} + \bar{n}_{12}\bar{n}_{23} + \bar{n}_{13}\bar{n}_{23},$$

and

$$\Delta = \bar{n}_{12}\bar{n}_{13}\bar{n}_{14} + \bar{n}_{12}\bar{n}_{13}\bar{n}_{24} + \bar{n}_{12}\bar{n}_{13}\bar{n}_{34} + \bar{n}_{12}\bar{n}_{14}\bar{n}_{23}$$
$$+ \bar{n}_{12}\bar{n}_{14}\bar{n}_{34} + \bar{n}_{12}\bar{n}_{23}\bar{n}_{24} + \bar{n}_{12}\bar{n}_{23}\bar{n}_{34} + \bar{n}_{12}\bar{n}_{24}\bar{n}_{34}$$
$$+ \bar{n}_{13}\bar{n}_{14}\bar{n}_{23} + \bar{n}_{13}\bar{n}_{14}\bar{n}_{24} + \bar{n}_{13}\bar{n}_{23}\bar{n}_{24} + \bar{n}_{13}\bar{n}_{23}\bar{n}_{34}$$
$$+ \bar{n}_{13}\bar{n}_{23}\bar{n}_{34} + \bar{n}_{13}\bar{n}_{24}\bar{n}_{34} + \bar{n}_{14}\bar{n}_{23}\bar{n}_{24} + \bar{n}_{14}\bar{n}_{24}\bar{n}_{34}.$$

$$(6.6)$$

For the data of Example 6.1, we are concerned with testing the hypothesis:

$$H_0 : \pi_{1.} = \pi_{.1}, \quad \pi_{2.} = \pi_{.2}, \quad \pi_{3.} = \pi_{.3}, \quad \pi_{4.} = \pi_{.4},$$
$$d_1 = -40, \quad d_2 = 10, \quad d_3 = 24, \quad d_4 = 6$$
$$a_1 = 83.75, \quad a_2 = 41.5, \quad a_3 = 111.5, \quad a_4 = 299.$$

TABLE 6.5

CJR Assessment versus CASS Assessment

CASS	Mild	Moderate	Severe	Very Severe	Total
Mild	11	0	0	0	11
Moderate	15	18	6	0	39
Severe	2	4	3	0	9
Very severe	0	0	0	1	1
Total	28	22	9	1	60

Note: The hypothesis of marginal homogeneity is rejected, $X^2_{(4)}$ = 17.4 (*p*-value = 0.0079).

The value of X^2_4 is 46.7 (*p*-value < 0.0001). We conclude that the hypothesis of absence of interrater bias is not supported by the data.

Example 6.2

The data presented in Tables 6.5 and 6.6 are the result of a study aiming at evaluating agreement between a clinical asthma severity score (CASS) and clinical judgment rating (CJR) made by pediatricians. The categories of assessments describe the severity grades: (middle, moderate, severe, and very severe).

Note from Table 6.5, that the "very severe" category has only one subject and therefore, it might be tempting to collapse the "very severe" category with the "severe category." The results are summarized in Table 6.7.

On comparing the analyses in Tables 6.7 and 6.8 we note that the SE has increased, and the value of the estimated coefficient of agreement has dropped, after collapsing the last two categories. Bartfay and Donner (2000) addressed this

TABLE 6.6

Data Analysis for the Agreement between the CASS and CRJ

Kappa	Value	SE	95% Lower Limits	95% Upper Limits
Simple kappa	0.3112	0.0874	0.1400	0.4825
Weighted kappa	0.3878	0.0917	0.2079	0.5676

TABLE 6.7

Cross Classification Table of CASS and CJR Classification

CASS	Mild	Moderate	Severe + Very Severe	Total
Mild	11	0	0	11
Moderate	15	18	6	39
Severe + very severe	2	4	4	10
Total	28	22	10	60

Note: Again, the marginal homogeneity test has a $X^2_{(3)}$ = 17.4 , and a *p*-value = 0.0006.

TABLE 6.8

Data Analysis of Table 6.7

Kappa	Value	SE	95% Lower Limits	95% Upper Limits
Simple kappa	0.3059	0.0869	0.1357	0.4762
Weighted kappa	0.3612	0.0859	0.1930	0.5295

issue in details, and demonstrated through extensive simulations that there are clear advantages to not collapsing the categorical data into lower dimensional tables. Moreover, the penalty may be quite high if the higher dimensional table is further collapsed into a 2 × 2 table. For example, if the three categories are collapsed into Category 1 = severe + very severe; Category 2 = otherwise, Tables 6.7 and 6.8 result in Table 6.9.

In this 2 × 2 table, $\hat{k} = 0.28$, $SE(\hat{k}) = 0.1569$, and the 95% confidence interval is (−0.0275, 0.5875). The $\hat{k}$ from the collapsed Table 6.9 is smaller and the confidence interval is much wider than the cases of 4 × 4 and 3 × 3 tables.

6.3 Agreement for Multiple Raters and Dichotomous Classification

6.3.1 Exchangeable Ratings

In this section, we consider the concept of agreement in a situation where different subjects are rated by different raters and the number of raters per subject varies. We focus on the case of dichotomous ratings. Fleiss and Cuzick (1979) provided an example where subjects may be hospitalized mental patients, the studied characteristic may be the presence or absence of schizophrenia, and the raters may be those psychiatry residents, out of a much larger pool, who happen to be on call when a patient is newly admitted. Not only may the particular psychiatrist responsible for one patient be different from those responsible for another, but different numbers of psychiatrists may provide diagnoses on different patients.

TABLE 6.9

Collapsed 2 × 2 Table

CASS	CJR Categories		Total
	1	2	
1	4	6	10
2	6	44	50
Total	10	50	60

TABLE 6.10

Data Layout for Multiple Raters and Two Categories

	Subject				
	1	2	3	...	K
Number of positive ratings	y_1	y_2	y_3		y_k
Number of raters	n_1	n_2	n_3		n_k

For convenience, we shall change the notation used to denote the number of subjects. Let k denote the number of subjects under study, n_i the number of raters rating the ith subject. Let y_{ij} be the rating reported by the jth rater on the ith subject, where $y_{ij} = 1$ if condition is present and $0 =$ else, and $y_i = \sum_{j=1}^{n_i} y_{ij}$ be the number of positive ratings on the ith subject. The data layout is shown in Table 6.10.

When $n_1 = n_2 = K = n_k$, Fleiss (1971) defined the estimators

$$\hat{P}_o = 1 - \frac{2}{k} \sum_{i=1}^{k} \frac{y_i(n - y_i)}{n(n-1)},$$

$$\hat{P}_e = 1 - 2\hat{\pi}(1 - \hat{\pi}),$$

where

$$\hat{\pi} = \sum_{i=1}^{k} \frac{y_i}{nk},$$

as estimates of crude agreement, chance agreement, and the probability of positive ratings, respectively. Fleiss (1971) showed that

$$\hat{k} = \frac{\hat{P}_o - \hat{P}_e}{1 - \hat{P}_e} = 1 - \frac{\sum_{i=1}^{k} y_i(n - y_i)}{kn(n-1)\hat{\pi}(1 - \hat{\pi})} = 1 - \frac{\hat{\pi} - \hat{P}_o}{\hat{\pi}(1 - \hat{\pi})}. \tag{6.7}$$

For variable number of ratings per subject, Fleiss and Cuzick (1979) extended the definition of κ_f to

$$\hat{\kappa}_f = 1 - \frac{1}{k(\bar{n} - 1)\hat{\pi}(1 - \hat{\pi})} \sum_{i=1}^{k} \frac{y_i(n_i - y_i)}{n_i},$$

where $\bar{n} = (1/k)\sum_{i=1}^{k} n_i$ is the average number of rating per subject, $\hat{\pi} = \sum_{i=1}^{k} y_i / \sum_{i=1}^{k} n_i$, and k is the number of subjects.

Lipsitz et al. (1994) proved the equivalence of $\hat{\kappa}_f$ an estimator obtained from the generalized estimating equation (GEE). They derived an asymptotically consistent estimator of $\hat{\kappa}_f$ as

$$\text{var}(\hat{\kappa}_f) \cong \sum_{i=1}^{k} \widehat{U}_i^2, \tag{6.8}$$

where

$$\widehat{U}_i = \frac{1}{k\hat{\pi}(1-\hat{\pi})}\left[\frac{y_i(n_i-y_i)}{n_i(\bar{n}-1)} - (\hat{\pi}-\bar{P}_{oo})\frac{n_i(n_i-1)}{n_i(\bar{n}-1)} - \frac{(1-\hat{\kappa}_f)(1-2\hat{\pi})}{\bar{n}}(y_i - n_i\hat{\pi})\right]$$

and

$$\hat{P}_{oo} = \frac{\sum_{i=1}^{k} y_i(y_i-1)}{\sum_{i=1}^{k} n_i(n_i-1)},$$

Fleiss and Cuzick (1979) showed that as $k \to \infty$, the estimator of Cohen's kappa in this situation is asymptotically equivalent to the estimated intraclass correlation $\hat{\rho}$ obtained from applying the within and between sums of squares formulae in the one-way random effects model to the y_{ij} (see Chapter 2). The measure $\hat{\rho}$ is given by

$$\hat{\rho} = \frac{\text{MSB} - \text{MSW}}{\text{MSB} + (n_0 - 1)\text{MSW}}, \tag{6.9}$$

where

$$\text{MSB} = \frac{1}{k-1}\left[\sum_i \frac{y_i^2}{n_i} - \frac{(\Sigma y_i)^2}{N}\right],$$

$$\text{MSW} = \frac{1}{N-k}\left[\sum_i y_i - \sum_i \frac{y_i^2}{n_i}\right],$$

$$n_0 = \frac{1}{k-1}\left[N - \frac{\sum_i n_i^2}{N}\right],$$

and $N = \Sigma n_i$. An estimator of the asymptotic variance for $\hat{\rho}$ was given by Mak (1988) as

$$\hat{\text{var}}(\hat{\rho}) = \frac{1}{k}[V_{11}C_1^2 + 2C_1C_2V_{12} + V_{22}C_2^2], \qquad (6.10)$$

where

$$C_1 = \frac{1}{(\bar{n} - 1)\hat{\pi}(1 - \hat{\pi})},$$

$$C_2 = -\frac{1 + (\bar{n} - 1)\left[\hat{\rho} + 2\hat{\pi}(1 - \hat{\rho})\right]}{\bar{n}(\bar{n} - 1)\hat{\pi}(1 - \hat{\pi})},$$

$$V_{11} = \frac{1}{k}\sum_{i=1}^{k}\left[\frac{y_i^4}{n_i^2} - f_i^2\right],$$

$$V_{12} = \frac{1}{k}\sum_{i=1}^{k}\left[\frac{y_i^3}{n_i} - \hat{\pi}n_i f_i\right],$$

$$V_{22} = \frac{1}{k}\sum_{i=1}^{k}\left[y_i^2 - n_i^2\hat{\pi}^2\right],$$

and

$$f_i = \hat{\pi}(1 - \hat{\pi})\left[1 + (n_i - 1)\hat{\rho}\right] + n_i\hat{\pi}^2.$$

6.3.2 Test for Interrater Bias

In Section 6.2, we assumed that each of the samples of subjects is assigned to one of c categories by one rater and to one of the same c categories by another rater. A comparison of the two resulting marginal distributions was made by means of Stuart and Maxwell's generalization of McNemar's test. Similarly, in the case of multiple raters, one may be interested in assessing the extent of homogeneity among of percentages of positive ratings done by the raters. To clarify the idea, we consider the following example from Shoukri and Pause (1999).

As a part of a "problem based learning" program, senior undergraduate students at the Ontario Veterinary College were asked to identify (from x-rays) foals with cervical vertebral malformation (CVM). For students participated in the exercise were asked to independently classify each of 20 x-rays as affected ("1") or not ("0"). The data are given in Table 6.11.

TABLE 6.11

Assessments of X-Ray by Four Students for Identification
of CVM Foals

X-ray	A	B	C	D	Total
			Raters		
1	0	0	0	0	0
2	0	0	1	0	1
3	1	1	1	1	4
4	1	1	1	1	4
5	1	1	0	1	3
6	0	0	0	0	0
7	1	0	0	0	1
8	0	0	0	0	0
9	1	1	1	1	4
10	1	0	1	1	3
11	1	1	1	1	4
12	1	1	0	1	3
13	1	1	0	0	2
14	1	0	1	0	2
15	1	0	0	0	1
16	0	0	1	0	1
17	1	1	1	1	4
18	1	0	0	0	1
19	1	1	1	1	4
20	1	1	1	1	4
Total	15	10	11	10	46

Clearly, the marginal totals indicate the differences in the classification probabilities of the 4 student raters for the same set of x-rays. Testing for the significance of the differences among these marginal probabilities (or testing for interrater bias) can be done using Cochran's Q statistic. Let y_{ij} denote the score made by the jth rater on the ith subject ($i = 1, 2, ..., k, j = 1, 2, ..., n$) where $y_{ij} = 1$ if the ith subject is judged by the jth rater to have the condition, and as 0 otherwise let $y_{i.}$ be the total number of raters who judge the ith subject as a case, and let $y_{.j}$ be the total number of subjects judged by the jth clinician to be cases. Cochran's Q-statistic is given by

$$Q = \frac{n(n-1)\sum_{i=1}^{n}(y_{i.} - (y_{..}/n))^2}{ny_{..} - \sum_{j=1}^{k}y_{.j}^2},$$ (6.11)

where

$$y_{..} = \sum_{j=1}^{n} y_{.j}.$$

TABLE 6.12

General Data Layout for Multiple Rates and Dichotomous
Classification

Rater	Subject				
	1	2	...	k	Total
1	y_{11}	y_{12}		y_{1k}	$y_{1.}$
	y_{21}	y_{22}		y_{2k}	$y_{2.}$
$\vdots$					
n	y_{n1}	y_{n2}		y_{nk}	$y_{n.}$
Total	$y_{.1}$	$y_{.2}$		$y_{.k}$	$y_{..}$

The general data layout is given in Table 6.12. Under the null hypothesis of absence of interrater bias, Q is approximately distributed as chi-square with $n-1$ degrees of freedom.

Note that in the case of two clinicians (i.e., $n=2$) Cochran's Q statistic is equivalent to McNemar's test.

Example 6.3

For the CVM data of Table 6.10, $Q = 6.375$ with three degrees of freedom. This leads us to declare the absence of interrater bias at $\alpha = 0.05$.

The result of applying the one-way ANOVA to get an estimate of ρ (reliability kappa) as in Chapter 2 can be obtained from the following SAS code listed below.

```
proc freq data=cvm;
tables A*B / agree;
run;
proc freq data=cvm;
tables A*C / agree;
run;
proc freq data=cvm;
tables A*D / agree;
run;
proc freq data=cvm;
tables B*C / agree;
run;
proc freq data=cvm;
tables B*D / agree;
run;
proc freq data=cvm;
tables C*D / agree;
run;

/* Calculating ICC from random effects model */
data new;
set CVM;
rater='rater_A';
```

```
score=A;
output;
rater='rater_B';
score=B;
output;
rater='rater_C';
score=C;
output;
rater='rater_D';
score=D;
output;
drop A B C D;
run;


/* Two way random effects model */
proc GLM data=new;
class X_ray rater;
model score=X_ray rater;
random X_ray rater;
run;


/* one Way random effect model*/
proc GLM data=new;
class X_ray;
model score=X_ray;
random X_ray;
run;


/* Testin for interrater bias */
proc print data=new;
run;
  proc sort data=new;
  by rater;
proc freq data=new;
tables X_ray*rater*score /cmh;
run;
```

We constructed a 2×2 table for all pairs of raters. The pairwise Cohen's kappa coefficients of agreement are summarized in Table 6.13.

As can be seen the lowest agreement is between raters (A,B), and the highest agreement is between (B,D). A simple average of the above coefficients gives $\tilde{\kappa}_{ave} = 0.46$. The ANOVA estimator of the overall ICC can be shown to be very close to the simple average of the pairwise agreement. Under a two-way random effects model, we found that the components of variance of x-ray, rater, and within x-ray are given respectively by $\hat{\sigma}_b^2 = 0.1207$, $\hat{\sigma}_r^2 = 0.0079$, and, $\hat{\sigma}_w^2 = 0.125$, and hence, $\hat{\rho}_2 = (0.1207/0.1207 + 0.0079 + 0.125) = 0.476$.

Note that the component of variance due to rater (rater effect) is quite small, and may be ignored. When a one-way random effects model

TABLE 6.13

Pairwise Agreement Using Cohen's Kappa

A-B	A-C	A-D	B-C	B-D	C-D
0.16	0.50	0.50	0.30	0.80	0.50

ANOVA model is fitted, we get $\hat{\sigma}_b^2 = 0.118$, $\hat{\sigma}_b^2 = 0.133$, and hence, $\hat{\rho}_1 = (0.118/0.118 + 0.133) = 0.469$. Apparently $(\bar{\kappa}_{ave}, \hat{\rho}_1, \hat{\rho}_2)$ give almost similar levels of agreement.

6.4 Probability Models

6.4.1 Bahadur's Representation

For the case of two raters and dichotomous classification, we are able to give the index of agreement kappa a specific feature, in that the classification probabilities are expressed as functions of both kappa and π. The model is called "common correlation model" (CCM) because the intraclass correlation parameter ρ or (reliability kappa) is assumed fixed across subjects. The extension to the situation of multiple raters and (0, 1) categories requires a generalization of the CCM.

Bahadur (1961) suggested that the joint probability distribution of the vector of binary responses $(y_{i1}, y_{i2}, ..., y_{in})$ from the ith subject may be written as

$$f(y_{i1}, ... y_{in}) = \prod_{j=1}^{n} \pi^{y_{i1}}(1-\pi)^{1-y_{i1}}$$

$$\times \left[1 + \rho \sum_{k<l} w_{ik}w_{il} + \rho_3 \sum_{k<l<m} w_{ik}w_{il}w_{im} + ... \rho_n \Sigma w_{i1}w_{i2} ... w_{in} \right], \quad (6.12)$$

where

$$w_{ij} = \left(y_{ij} - \pi / \{\pi(1-\pi)\}^{1/2} \right), \quad \rho_2 = \rho = E(w_{ik}w_{il}), \quad \rho_n = E(w_{i1}w_{i2} ... w_{in}).$$

For interrater agreement, the parameter of interest is the second-order correlation ρ, which may be shown to be equivalent to the ICC as obtained from the one-way random effects model (Mekibib et al., 2001). For the case of three raters, $(n = 3)$ the Bahadur's model requires the specification of the parameter ρ_3 as the "three-way association parameter." But ρ_3 does not have a readily

TABLE 6.14

Data Layout for Three Raters and Dichotomous Categories under Bahadur's Probabilistic Representation

Category	Ratings	Frequency	Probability
0	(0, 0, 0)	m_0	$P(0)$
1	(0, 0, 1), (0, 1, 0), (1, 0, 0)	m_1	$P(1)$
2	(1, 1, 0), (1, 0, 1) (0, 1, 1)	m_2	$P(2)$
3	(1, 1, 1)	m_3	$P(3)$

available interpretation in terms of agreement; in fact, in this model, it is considered a nuisance parameter. Prentice (1988) suggested expressing ρ_3 in terms of π and ρ as given by

$$\rho_3 = \rho \left\{ \left(\frac{1-\pi}{\pi} \right)^{1/2} - \left(\frac{\pi}{1-\pi} \right)^{1/2} \right\}.$$

For this special situation, we summarize the data layout in Table 6.14. where

$$P(0) = P(Y_i = 0) = (1-\pi)^3 + \rho\pi\left((1-\pi)^2 + (1-\pi)\right),$$

$$P(1) = P(Y_i = 1) = 3\pi(1-\pi)^2(1-\rho),$$

$$P(2) = P(Y_i = 2) = 3\pi^2(1-\pi)(1-\rho),$$

$$P(3) = P(Y_i = 3) = \pi^3 + \rho\pi\left(1-\pi^2\right).$$

$Y_i = Y_{i1} + Y_{i2} + Y_{i3}$, and m_j = number of subjects whose total score Y_i is $j (j = 0, 1, 2, 3)$ and $i = 1, 2, ..., k$.

For example, if we have six subjects, such that

Total score for subject 1 = 0
Total score for subject 2 = 1
Total score for subject 3 = 1
Total score for subject 4 = 3
Total score for subject 5 = 2
Total score for subject 6 = 0

then, $m_0 = 2$, $m_1 = 2$, $m_2 = 1$, $m_3 = 1$.

We note that the estimator $\hat{\rho}$ of interrater agreement is the same as in the previous section. Its variance, for $n = 3$, was concisely given by Mekibib et al. (2001) as

$$\text{var}(\hat{\rho}) = \frac{1-\rho}{3k}\left[(1 - 8\rho)(1 - \rho) + \frac{\rho(3 - 2\rho)}{\pi(1 - \pi)}\right]. \tag{6.13}$$

A consistent estimator of $\text{var}(\hat{\rho})$ may be obtained by replacing ρ by $\hat{\rho}$ and π by $\hat{\pi} = (3m_3 + 2m_2 + m_1/3\kappa)$ in Equation 6.13.

A $100(1 - \alpha)\%$ confidence interval can be constructed as $\hat{\rho} \pm Z_{1-\alpha/2}\sqrt{\text{var}(\hat{\rho})}$, where $Z_{1-\alpha/2}$ is the $100(1 - \alpha/2)$ percentile point of the standard normal distribution.

It should be noted that the form of Bahadur's model used in this section assumes no rater bias, that is, each rater is characterized by the same underlying success rate π. As discussed by Landis and Koch (1977), the estimator $\hat{\rho}$ and the corresponding confidence interval are most appropriate when the emphasis of the study is directed to the measurement process itself rather than the potential differences among raters.

Example 6.4

Each of 20 patients has been assessed independently by three raters ($n = 3$) for the presence or absence of a disease condition. Assuming exchangeability of ratings, the total positive rating in the three assessment is denoted by x. The data are given in Table 6.15.

Here, $m_0 = 5$, $m_1 = 6$, $m_2 = 8$,

TABLE 6.15

Classification of 20 Subjects by Three
Raters into Two Categories

Patient	x	Patient	x
1	2	11	3
2	2	12	2
3	0	13	1
4	1	14	1
5	0	15	2
6	0	16	2
7	1	17	1
8	0	18	2
9	1	19	2
10	2	20	0

Therefore, $\hat{\pi} = (3 + 16 + 6/60) - 0.417$, $\hat{\rho} = 0.056$.

$$\text{var}(\hat{\rho}) = \frac{0.944}{60}\left[(0.552)(0.944) + \frac{(0.056)(2.888)}{(0.417)(1.417)}\right]$$

$$= 0.186.$$

Hence, $SE(\hat{\rho}) = 0.137$.

6.4.2 Dirichlet Multinomial Model: Three Categories

6.4.2.1 Dirichlet Trinomial Model

In this section, we shall review alternative probabilistic models developed to facilitate sensible statistical inference on categorical agreement. We first assume that we have n subjects being classified into one of three mutually exclusive categories by two randomly selected raters. Let X_1, X_2 and X_3 be the number of pairs of ratings that belong to categories 1, 2, and 3, and let $P_1, P_2, P_3 = (1 - P_1 - P_2)$ be the probability of belonging to each category. The joint distribution of $(X_1, X_2, X_3)'$ is that of a multinomial so that conditional on $P = (P_1, P_2, P_3)'$

$$\Pr\left(X_1 = x_1, X_2 = x_2, X_3 = x_3 \middle| P\right) = \frac{2!}{\prod_{i=1}^{3} x_i!}\prod_{i=1}^{3} P_i^{x_i}.$$

Here $x_1 + x_2 + x_3 = 2$, and $P_1 + P_2 + P_3 = 1$. Following Bartfay and Donner (2001), to affect the correlation between the pairs of ratings, we assume that $P = (P_1, P_2, P_3)'$ follows a Dirichlet distribution with parameters $(\alpha_1, \alpha_2, \alpha_3)$ and the corresponding density function

$$f(\underline{P}) = \frac{\Gamma(\alpha_1 + \alpha_2 + \alpha_3)}{\prod_{i=1}^{3}\Gamma(\alpha_i)}\prod_{i=1}^{3} P_i^{\alpha_i - 1}.$$

The unconditional distribution of $(X_1, X_2, X_3)'$ is therefore, given by

$$p(x_1, x_2, x_3) = \int_0^1\int_0^1\int_0^1 f(P)P(X_1 = x_1, X_2 = x_2, X_3 = x_3)\,dP_1\,dP_2\,dP_3$$

$$= 2\frac{\Gamma(\alpha_1 + \alpha_2 + \alpha_3)\Gamma(\alpha_1 + x_1)\Gamma(\alpha_2 + x_2)\Gamma(\alpha_3 + x_3)}{x_1!x_2!x_3!\Gamma(\alpha_1)\Gamma(\alpha_2)(\alpha_3)\Gamma(2 + \alpha_1 + \alpha_2 + \alpha_3)}.$$

This is called the Dirichlet trinomial model (DTM). Under the transformations

$$\kappa = \left(\alpha_1 + \alpha_2 + \alpha_3 + 1\right)^{-1},$$

$$\pi_j = \alpha_j / \left(\alpha_1 + \alpha_2 + \alpha_3\right) \quad j = 1, 2,$$

$$\pi_3 = 1 - \pi_1 - \pi_2.$$

The cross classification 3×3 table can be written as in Table 6.16.

Because of the symmetry, Table 6.17 may be collapsed into the following categorical structure.

Therefore, n_1, n_2, n_3 are the frequencies of concordant pairs of ratings, while n_4, n_5, n_6 are the frequencies of the discordant pairs. From Table 6.13, the crude agreement is given by

$$P_o = \pi_1^2 + \pi_2^2 + \pi_3^2 + \kappa \left[\pi_1 (1 - \pi_1) + \pi_2 (1 - \pi_2) + \pi_3 (1 - \pi_3) \right],$$

and the chance agreement P_e is obtained by setting $\kappa = 0$, so that

$$P_e = \pi_1^2 + \pi_2^2 + \pi_3^2.$$

It can therefore, be established that $\kappa = (P_0 - P_e / 1 - P_e)$ is the chance corrected agreement.

Following Fleiss (1981) and Bartfay and Donner (2001), the moment estimators of the model parameters are given as

$$\hat{\pi}_1 = \frac{(2n_1 + n_4 + n_5)}{2n},$$

$$\hat{\pi}_2 = \frac{(2n_2 + n_4 + n_6)}{2n},$$

$$\hat{\pi}_3 = \frac{(2n_3 + n_5 + n_6)}{2n},$$

TABLE 6.16

Cross Classification Probabilities According to the DTM

Rater (2) Categories	Rater (1) Categories			Total
	1	2	3	
1	$\pi_1^2 + \kappa_1 \pi_1 (1 - \pi_1)$	$\pi_1 \pi_2 (1 - \kappa)$	$\pi_1 \pi_3 (1 - \kappa)$	π_1
2		$\pi_2^2 + \kappa \pi_2 (1 - \pi_2)$	$\pi_2 \pi_3 (1 - \kappa)$	π_2
3			$\pi_3^2 + \kappa \pi_3 (1 - \pi_3)$	π_3
Total	π_1	π_2	π_3	1

TABLE 6.17

Collapsed Probabilities under the Assumption of Exchangeability

Category	Ratings	y	Frequency
1	(1, 1)	2	n_1
2	(2, 2)	4	n_2
3	(3, 3)	6	n_3
4	(1, 2) or (2, 1)	3	n_4
5	(1, 3) or (3, 1)	4	n_5
6	(2, 3) or (3, 2)	5	n_6
Total	1		n

$$\hat{\kappa} = 1 - \frac{n_D}{n\left(1 - \sum_{i=1}^{3}\hat{\pi}_i^2\right)}, \qquad (6.14)$$

and $n_D = n_4 + n_5 + n_6$ is the total number of discordant pairs.

Bartfay and Donner (2001) derived the asymptotic variance of $\hat{\kappa}$ and its estimator as

$$
\text{var}(\hat{\kappa}) = \frac{(1-\hat{\kappa})^2}{nD^2}\left[4\left\{ (1-\hat{\kappa})\sum_{i=1}^{3}\hat{\pi}_i^4 + \hat{\kappa}\sum_{i=1}^{3}\hat{\pi}_i^3 \right\} \right.
$$

$$
\left. + 2\left\{ \frac{T_1}{1-\hat{\kappa}} + 2T_2 + T_3(1-\hat{\kappa}) \right\} - (D+2)^2 \right] \qquad (6.15)
$$

where

$$D = \sum_{i=1}^{3}\hat{\pi}_i^2 - 1,$$

$$T_1 = \hat{\pi}_1\hat{\pi}_2 + \hat{\pi}_1\hat{\pi}_3 + \hat{\pi}_2\hat{\pi}_3,$$

$$T_2 = \hat{\pi}_1\hat{\pi}_2(\hat{\pi}_1 + \hat{\pi}_2) + \hat{\pi}_1\hat{\pi}_3(\hat{\pi}_1 + \hat{\pi}_3) + \hat{\pi}_2\hat{\pi}_3(\hat{\pi}_2 + \hat{\pi}_3),$$

$$T_3 = \hat{\pi}_1\hat{\pi}_2(\hat{\pi}_1 + \hat{\pi}_2)^2 + \hat{\pi}_1\hat{\pi}_3(\hat{\pi}_1 + \hat{\pi}_3)^2 + \hat{\pi}_2\hat{\pi}_3(\hat{\pi}_2 + \hat{\pi}_3)^2.$$

For large n or $(1 - \alpha)100\%$ confidence interval on κ was given by Bartfay and Donner (2001) as

$$\hat{\kappa} \pm Z_{1-\alpha/2}\sqrt{\text{var}(\hat{\kappa})}.$$

Through extensive Monte Carlo simulations they showed that the above confidence has a coverage probability close to the nominal one.

Example 6.5

Consider the data in Table 6.3. Let us use the following categorization Mild = 1, Moderate = 2, and Severe + Very Severe = 3. Hence, the data would have the representation shown in Table 6.18.

$$\hat{\pi}_1 = \frac{22 + 15 + 2}{120} = 0.325,$$

$$\hat{\pi}_2 = \frac{36 + 15 + 10}{120} = 0.508,$$

$$\hat{\pi}_3 = \frac{8 + 2 + 10}{120} = 0.167,$$

$$D = (0.325)^2 + (0.508)^2 + (0.167)^2 - 1$$

$$= 0.392 - 1$$

$$= -0.608,$$

$$\hat{\kappa} = 1 - \frac{27}{60(1 - 0.392)} = 0.26,$$

$$T_1 = 0.3042, \quad T_2 = 0.221, \quad T_3 = 0.166.$$

It will be left as an exercise to the reader to derive the asymptotic variance given in Equation 6.15.

TABLE 6.18

Cross Classification of CASS and CJR

Category	Rating	y	Frequency
1	(1,1)	2	$11 = n_1$
2	(2,2)	4	$18 = n_2$
3	(3,3)	6	$4 = n_3$
4	(1,2) or (2,1)	3	$15 = n_4$
5	(1,3) or (3,1)	4	$2 = n_5$
6	(2,3) or (3,2)	5	$10 = n_6$
			60

6.5 Multiple Raters and Multiple Categories

Fleiss (1971) proposed a generalization of Cohen's kappa statistic to the measurement of agreement among a constant number of raters n. Each of the k subjects is rated by $n > 2$ raters independently into one of c mutually exclusive and exhaustive nominal categories. The motivating example was a study in which each of 30 patients was rated by six psychiatrists (selected randomly from a pool of 43 psychiatrists) into one of five categories. Let k_{ij} be the number of raters who assign the ith subject to the jth category $(i = 1, 2, ..., k, j = 1, 2, ..., c)$, and define

$$p_j = \frac{1}{nk} \sum_{i=1}^{k} k_{ij}.$$

Here p_j is the proportion of all assignments to the jth category. The chance-corrected measure of overall agreement proposed by Fleiss (1971) is

$$\hat{k}_{mc} = \frac{\sum_{i=1}^{k} \sum_{j=1}^{c} k_{ij}^2 - kn\left\{1 + (n-1)\sum_{j=1}^{c} p_j^2\right\}}{kn(n-1)(1 - \sum_{j=1}^{c} p_j^2)} \qquad (6.16)$$

(the subscript mc is for "multiple categories")

In addition to the $\hat{k}_{mc}$ for measuring overall agreement, Fleiss (1971) proposed another statistic to measure the extent of agreement in assigning a subject to a particular category. His suggested measure of the beyond-chance agreement in assignment to category j is given by

$$\hat{k}_j = \frac{\sum_{i=1}^{k} k_{ij}^2 - knp_j\left\{1 + (n-1)p_j\right\}}{kn(n-1)p_j(1 - p_j)}. \qquad (6.17)$$

Clearly, $\hat{k}_{mc}$ in Equation 6.16 is the weighted average of $\hat{k}_j$ given in Equation 6.17. The corresponding weights $p_j(1 - p_j)$ when subjects are being rated by different numbers of raters. Landis and Koch (1977b) associated $\hat{k}_{mc}$ with the ICC computed for one-way random effects, ANOVA with the single factor corresponding to the random (subjects). Davies and Fleiss (1982) demonstrated this equivalence for a two-way balanced layout. They proposed a kappa-like statistic for a set of multinomial random variables arrayed in a two-way (subject by rater) layout. Furthermore, they showed that this proposed statistic may be obtained either via chance-correction of the average proportion of pairwise agreement, or via an analysis of variance for a two-way layout. Applications include the case where each of the same set of several clinicians classifies each of a sample of patients into one of several mutually exclusive categories.

Construction of confidence intervals on κ_{mc} has been difficult due to the fact that a variance of $\hat{\kappa}_{mc}$, even asymptotically, is not available. However, Davies and Fleiss (1982) provided an asymptotic variance for $\hat{\kappa}_{mc}$, only in the null case (i.e., when $\hat{\kappa}_{mc} = 0$). The authors discuss some interesting applications where the hypothesis that the population kappa equals zero might be of interest.

For computational ease, we write expression (Equation 6.16) as

$$\hat{\kappa}_{mc} = \frac{p_o - p_e}{1 - p_e},$$

$$p_o = \frac{\sum_{i=1}^{k} \sum_{j=1}^{c} k_{ij}^2 - nk}{kn(n-1)},$$

and

$$p_e = \sum_{j=1}^{c} p_j^2,$$

where

$$p_j = \frac{1}{nk} \sum_{i=1}^{k} k_{ij}.$$

The estimated variance of $\hat{\kappa}_{mc}$ was given by Woolson (1987) as

$$\text{Var}(\hat{\kappa}_{mc}) = \frac{2}{kn(n-1)} \left[\frac{p_e - (2n-3)p_e^2 + 2(n-1)\sum_{j=1}^{c} p_j^3}{(1 - p_e)^2} \right].$$

Example 6.6

The data provided by Williams (1976), though for different objectives, may be used as an example to demonstrate the evaluation of $\hat{\kappa}_{mc}$.

As part of their clinical laboratory quality evaluation program, the College of American Pathologists (CAP) conducts a proficiency testing program for syphilis serology. Table 6.19 represents an assignment of each of 28 syphilis serology specimens tested independently by four clinical laboratories using the FTA-ABS test. The assignments are to the three categories: nonreactive (NR), borderline (BL), and reactive (R).

TABLE 6.19

Report on 28 Specimens by Four Laboratories

Specimen	Laboratory			
	0	1	2	3
1	R	R	R	R
2	R	R	R	R
3	BL	NR	NR	NR
4	BL	NR	NR	NR
5	BL	NR	NR	NR
6	R	R	R	R
7	BL	NR	NR	NR
8	R	R	R	R
9	NR	NR	NR	NR
10	NR	NR	NR	NR
11	R	R	R	R
12	R	R	BL	BL
13	R	R	R	R
14	R	R	BL	BL
15	R	R	R	R
16	R	R	NR	BL
17	R	R	NR	BL
18	R	R	R	R
19	R	R	R	R
20	BL	BL	NR	NR
21	R	R	R	R
22	BL	NR	NR	NR
23	BL	NL	NR	NR
24	BL	NL	NR	NR
25	R	R	R	R
26	NR	NR	NR	NR
27	R	R	R	R
28	NR	NR	NR	NR

The evaluation of $\hat{\kappa}_{mc}$ proceeds as follows: First, we construct a table showing the number of assignments to each category (Table 6.20).

$$p_o = \frac{1}{(28)(4)(3)}[358 - (28)(4)] = 0.732.$$

Fleiss (1971) interpreted p_o as follows. Let a subject be selected at random and diagnosed by a randomly selected laboratory. If the subject were also diagnosed by a second randomly selected laboratory, the second diagnosis would agree with the first over 73% of the time.

TABLE 6.20

Values of k_{ij} for the Four Laboratories

(i) Specimen	NR (1)	BL (2)	R (3)	$\sum_{j=1}^{c} k_{ij}^2$
		(j)		
1			4	16
2			4	16
3	3	1		10
4	3	1		10
5	3	1		10
6			4	16
7	3	1		10
8			4	16
9	4			16
10	4			16
11			4	16
12		2	2	8
13			4	16
14		2	2	8
15			4	16
16	1	1	2	6
17	1	1	2	6
18			4	16
19			4	16
20	2	2		8
21			4	16
22	3	1		10
23	2	2		8
24	2	2		8
25			4	16
26	4			16
27			4	16
28	4			16
Total	39	17	56	358

$$p_1 = \frac{39}{(28)(4)} = 0.348, \quad p_2 = \frac{17}{(28)(4)} = 0.152, \quad p_3 = \frac{57}{(28)(4)} = 0.509.$$

$$p_e = \sum_{j=1}^{3} p_j^2 = (0.348)^2 + (0.152)^2 + (0.509)^2 = 0.403.$$

Therefore,

$$\hat{\kappa}_{mc} = \frac{0.732 - 0.403}{1 - 0.403} = 0.552.$$

Moreover,

$$\sum_{j=1}^{3} p_j^3 = (0.348)^3 + (0.152)^3 + (0.509)^3 = 0.1775.$$

Substituting in the variance expression we obtain

$$\text{var}(\hat{\kappa}_{mc}) = \frac{2}{28(4)(3)}\left[\frac{0.403 - 5(0.403)^2 + 2(2)(0.1775)}{(1 - 0.403)^2}\right] = 0.005.$$

An approximate 95% confidence interval on κ_{mc} can be constructed as $\hat{\kappa}_{mc} \pm 1.96\sqrt{\text{var}(\hat{\kappa}_{mc})}$. From the data this interval is given by (0.41, 0.69).
Clearly, the agreement between the laboratories is low. There are several reasons; the one that comes to mind is the possibility that the observer (laboratory) cannot distinguish, for some specimens, between the reactive and the nonreactive categories. The data may further be explored by examining the pairwise agreement. This will be left as an exercise to the reader.

6.6 Testing the Homogeneity of Kappa Statistic from Independent Studies

In a multicenter clinical trial, reliability studies are conducted independently in each of the several centers. This gives rise to several independent kappa statistics. Alternatively, they may rise from a single study in which subjects are divided into several strata, as discussed by Barlow et al. (1991). The main purpose of the investigation would be testing whether the level of interrater agreement as measured by the kappa statistics, can be regarded as homogeneous across centers, that is, to test $H_0 : \kappa_1 = \kappa_2 = \cdots = \kappa_k$, where κ_h denotes the population value of kappa in study h.

Donner et al. (1996) have developed a method for testing the homogeneity of k independent kappas of the intraclass form. Their underlying model assumes that k independent studies, involving $n = \sum_{h=1}^{k} n_h$ subjects, have been completed, where each subject is given a dichotomous rating (yes/no) by each of two raters.

Furthermore, it is assumed that the marginal probability of classifying a subject as success is constant across raters in a particular study (say π_h); however, this probability may vary across the k studies. In other words, there is no rater bias within studies. Under these assumptions, the probabilities of joint responses within study h arise from a trinomial model (obtained by collapsing the two discordant cells into a single cell) and are given as

Both successes: $P_{1h}(\kappa_h) = \pi_h^2 + \pi_h(1 - \pi_h)\kappa_h.$

One success and one failure: $P_{2h}(\kappa_h) = 2\pi_h(1 - \pi_h)(1 - \kappa_h).$

Both failures: $P_{3h}(\kappa_h) = (1 - \pi_h)^2 + \pi_h(1 - \pi_h)\kappa_h.$

For the hth study, the MLE for π_h and κ_h are given, respectively by

$$\hat{\pi}_h = \frac{2n_{1h} + n_{2h}}{2n_h},$$

$$\hat{\kappa}_h = 1 - \frac{n_{2h}}{2n_h\hat{\pi}_h(1 - \hat{\pi}_h)},$$

where n_{1h} is the number of subjects in study h who received "Yes" ratings from both raters, n_{2h} is the number who received one "Yes" and one "No" rating, n_{3h} is the number who received "No" ratings from both raters, and $n_h = n_{1h} + n_{2h} + n_{3h}$. An overall measure of agreement among the studies is estimated by computing a weighted average of the individual $\hat{\kappa}_h$, yielding

$$\hat{\kappa} = \frac{\sum_{h=1}^{k} n_h\hat{\pi}_h(1 - \hat{\pi}_h)\hat{\kappa}_h}{\sum_{h=1}^{k} n_h\hat{\pi}_h(1 - \hat{\pi}_h)}.$$

To test $H_0 : \kappa_1 = \kappa_2 = \cdots = \kappa_k$, Donner et al. (1996) proposed a goodness-of-fit test based on the statistic

$$X_G^2 = \sum_{h=1}^{k}\sum_{l=1}^{3} \frac{\{n_{lh} - n_h\hat{P}_{lh}(\hat{\kappa})\}^2}{n_h\hat{P}_{lh}(\hat{\kappa})},$$

where $\hat{P}_{lh}(\hat{\kappa})$ is obtained by replacing π_h by $\hat{\pi}_h$ and κ_h by $\hat{\kappa}$ in $P_{lh}(\kappa_h), l = 1, 2, 3;$ $h = 1, 2, \ldots, k$. Under the null hypothesis, X_G^2 follows an approximate chi-square distribution with $k - 1$ degrees of freedom.

Donner et al. (1996) discussed another test of homogeneity of several kappas from independent studies. From Bloch and Kraemer (1989), the large sample variance of $\hat{\kappa}_h$ is given by

$$\widehat{\text{var}}(\hat{\kappa}_h) = \left(\frac{1 - \hat{\kappa}_h}{n_h}\right)\left[(1 - \hat{\kappa}_h)(1 - 2\hat{\kappa}_h) + \frac{\hat{\kappa}_h(2 - \hat{\kappa}_h)}{2\hat{\pi}_h(1 - \hat{\pi}_h)}\right],$$

$$\text{Let } \hat{W}_h = \left(\widehat{\text{var}}\left(\hat{\kappa}_h\right)\right)^{-1}, \quad \text{and} \quad \tilde{\kappa} = \frac{\left(\sum_{h=1}^{k} \hat{W}_h \hat{\kappa}_h\right)}{\left(\sum_{h=1}^{k} \hat{W}_h\right)}.$$

An approximate test on the hypothesis of homogeneity is obtained by referring

$$X_V^2 = \sum_{h=1}^{k} \hat{W}_n \left(\hat{\kappa}_h - \tilde{\kappa}\right)^2$$

to tables of the chi-square distribution with $k - 1$ degrees of freedom.

Note that if $\hat{\kappa}_h = 1$ for any h, $\hat{W}_h$ is undefined. This event may occur frequently in samples of small and moderate size, and therefore, X_V^2 cannot be used. In contrast, the X_G^2 can be calculated except in the rare event when $\hat{\kappa}_h = 1$ for all $h = 1, 2, ..., \kappa$. Monte Carlo experiments designed by Donner et al. showed that both the statistics have similar properties for large samples ($n_h > 100$ for all, h).

In this case differences in power tend to be negligible except in the case of unequal π_h's or extreme unbalance in study sizes, where X_G^2 tends to have a small but consistent advantage over X_V^2. In general, their simulations showed that the X_G^2 goodness-of-fit statistic is preferable.

One of the limitations of the CCM used in this section is the assumption that each rater in a given study may be characterized by the same underlying success rate. However, we emphasize what Landis and Koch (1977) have noted, that this model is most appropriate when the main emphasis is directed at the reliability of the measurement process rather than in potential differences among raters. Similarly, Hale and Fleiss (1993) noted that the CCM permits measurement of agreement in many different settings: the reliability of a single rater based on independent replicate assignments by that rater; or the reproducibility of evaluations made by members of a pool of raters when different pairs of raters are selected from that pool.

When the assumption of a common success rate across raters within a study is not justifiable, methods using Cohen's kappa, as described by Fleiss (1981) and Fleiss and Cicchetti (1978) are appropriate. The assumption of a common rater effect (absence of interrater bias) in a particular study can be formerly tested by applying McNemar's test for homogeneity of marginal probabilities.

Example 6.7

Barlow et al. (1991) provided the data in the following table, from a randomized clinical trial to test the effectiveness of silicone fluid versus gas in the management

TABLE 6.21

Agreement between Ophthalmologist and Reading Center Classifying
Superior Nasal Retinal Breaks Stratified by PVR Grade

	PVR Grade				
Rating	C_3	D_1	D_2	D_3	Total
(1, 1)	1	6	5	3	15
(1, 0) or (0, 1)	9	8	11	9	37
(0, 0)	65	46	54	33	198
Total	75	60	70	45	250

of proliferative vitreoretinopathy (PVR) of vitrectomy. The main interest was the
degree of agreement on the presence or absence of retinal breaks in the superior nasal quadrant as ascertained clinically by the operating ophthalmic surgeon
and photographically by an independent fundus photograph reading center. The
subjects in the study were stratified by PVR grade which measures the severity of
disease measured on a continuum of increasing pathology graded as C3, D1, D2,
or D3. The data are presented in Table 6.21 with stratification by PVR grade. The
hypothesis was that knowledge of the PVR grade might influence the observers'
marginal probabilities of noting retinal breaks. For the purpose of illustration we
use the data to demonstrate the homogeneity of measures of agreement across
PVR grades.

Summary Measures from Table 6.21

PVR Grade	2 × 2 Table				n_h	$\hat{k}_h$	$\hat{\pi}$
C_3			1	0	75	0.117	0.073
	1		1	9	$n_{11} = 1,$	$n_{21} = 9,$	$n_{31} = 65$
	0		0	65			
D_3							
			6	5	60	0.52	0.167
			3	46	$n_{21} = 6,$	$n_{22} = 8,$	$n_{32} = 46$
D_2							
			5	9	70	0.38	1.50
			2	54	$n_{13} = 5,$	$n_{23} = 11,$	$n_{33} = 54$
D_3							
			3	7	45	0.28	0.167
			2	33	$n_{14} = 3,$	$n_{24} = 9,$	$n_{34} = 33$

The sample estimates $\hat{k}_h$, $h = 1, 2, 3, 4$ are given respectively, by 0.12, 0.52,
0.38, and 0.28 with respective sample sizes n_h, given by 75, 60, 70, and 45. For
testing $H_0 : \kappa_1 = \kappa_2 = \kappa_3 = \kappa_4$, the three degree of freedom chi-square goodness-of-
fit statistic is given by $X_C^2 = 2.97$ ($p = 0.369$), showing that there is no evidence
against the assumption that each stratum is characterized by the same value of
κ. The results obtained for applying the large-sample variance approached are

similar, yielding $X_W^2 = 3.49$ ($p = 0.322$). An estimate of the common κ is given by $\hat{\kappa} = 0.35$.

6.6.1 Combined Estimator of Kappa from Several Independent Studies: A General Approach

In the previous section, we discussed a method of combining and testing equality of several kappa parameters in the special case of two rates and two categories. The main objective was to account possible confounder by stratifying on the levels of one possible confounder.

In this section, we discuss situations where several estimates of the kappa coefficient are obtained from similar studies. We shall distinguish between two approaches; the first is called the fixed effects approach, and the other is the random effects approach. The methodology for both approaches is adopted from methodologies of meta-analysis of independent studies.

6.6.2 Fixed Model Approach

Failing to stratify by the pathology grade as in Example 6.4 (see Table 6.21) may produce a misleading estimate of the coefficient to agreement.

Suppose that we wish to assess the overall interrater agreement of two raters ($i = 1$, 2) with classification of n_h subjects per stratum or study $h(h = 1, 2, ..., k)$ into one of c categories ($j = 1, 2, ..., C$). We use the following $c \times c$ table to summarize the data structure for a given stratum (study). For completeness of the presentation, and as an example for the care of two raters and multiple categories, we summarize out notations in Table 6.22.

As before, $0 < \pi_{ijk} < 1$ denotes the cell probability, $\pi_{i\cdot k} = \sum_{j=1}^{c} \pi_{ij.k}$, and $\pi_{\cdot jk} = \sum_{i=1}^{c} \pi_{ijk}$ for $i = 1$, 2, and $j = 1, 2, ..., c$. Let n_{ijk} denote the corresponding cell frequency of the cell probability π_{ijk}, where $n_k = \sum_i \sum_j n_{ijk}$.

Define $\hat{\pi}_{ijk} = n_{ijk}/n_k$, $\hat{\pi}_{i\cdot k} = \sum_{j=1}^{c} \hat{\pi}_{ijk}$, and $\hat{\pi}_{\cdot jk} = \sum_{i=1}^{c} \hat{\pi}_{ijk}$. Conditional on study k, let $\hat{k}_k = \left(\hat{P}_{ok} - \hat{P}_{ek}\right)/\left(1 - \hat{P}_{ek}\right)$, where $\hat{P}_{ok} = \sum_{j=1}^{c} \hat{\pi}_{jjk}$, $\hat{P}_{ek} = \sum_{i=1}^{c} \hat{\pi}_{i\cdot k}\hat{\pi}_{\cdot ik}$.

TABLE 6.22

Classification by Two Raters into Categories for the kth Stratum

Rater (2) Category	Rater (1) Category				
	1	2	...	C	Total
1	π_{11k}	π_{12k}		π_{1ck}	$\pi_{1.k}$
2	π_{21k}	π_{22k}		π_{2ck}	$\pi_{2.k}$
⋮	⋮				
c	π_{c1k}	π_{c2k}		π_{cck}	$\pi_{c.k}$
Total	$\pi_{.1k}$	$\pi_{.2k}$		$\pi_{.ck}$	1

For large n_k, let the asymptotic variance of $\hat{k}_k$ be denoted by $\hat{v}_k$ $(k = 1, 2, \ldots, k)$. In Section 6.6, we considered a summary estimator

$$\hat{\kappa}_c = \frac{\left(\sum_{k=1}^{k} \hat{W}_k \hat{\kappa}_k\right)}{\sum_{k=1}^{k} \hat{W}_k}$$

with optimal weights $\hat{W}_k = \hat{v}_k^{-1}$, for the 2×2 case. It is easily shown that for the general case the combined estimator $\hat{\kappa}_c$ has asymptotic variance $\hat{v}_k = 1/\sum_{k=1}^{k} \hat{W}_k$. When we substitute the unknown parameters by their sample estimates we obtain estimates of the variances, and hence, the variance of the combined estimate. The suggested $100(1 - \alpha)\%$ confidence interval of the underlying common k for the k studies is

$$\sum_{k=1}^{k} \hat{W}_k \hat{\kappa}_k \pm z_{\alpha/2} \sqrt{\left(\sum_{k=1}^{k} \hat{W}_k\right)^{-1}}.$$

6.6.3 Meta-Analysis Approach (Random Effects Model)

Let us assume that the estimated value of agreement from the lth study is $\hat{\kappa}_l$, and assume a true value of this estimate is κ_l. A general model is then specified as $\hat{\kappa}_l | \kappa_l \sim N(\kappa_l, \eta_l)$ and $\kappa_l \sim N(\kappa, \tau^2)$. Under this set up, κ is the overall agreement coefficient. The variance η_l^2, for $l = 1, 2, \ldots, k$ are within-study or the study-specific variance, and τ^2 is the across-studies variance. In the meta-analysis (MA) literature (see Biggerstaff and Tweedie, 1997; Hardy and Thompson, 1998) the variance τ^2 is often referred to as the heterogeneity variance within the random effects MA framework. Under the assumption of independence of studies, we note that $\text{var}(\hat{\kappa}_l) = \eta_l^2 + \tau^2 = \tau^2(1 + r_l)$ for $l = 1, 2, \ldots, k$, where $r_l = \eta_l^2/\tau^2$, $\tau^2 \neq 0$. Since η_l^2 can be estimated from the data by $\hat{\eta}_l^2$, one needs to find an estimate for τ^2.

Alternatively, Sidik and Jonkman (2005) suggested, as crude or a first stage estimator for τ^2 as

$$\hat{\tau}_o^2 = \frac{1}{k \sum_{l=1}^{k} (\hat{\kappa}_l - \bar{\kappa})^2},$$

where $\bar{\kappa} = 1/\kappa \sum_{l=1}^{k} \hat{\kappa}_l$. Therefore, r_l may be estimated by using $\hat{r}_l = \hat{\eta}_l^2/\tau_o^2$ for $l = 1, 2, \ldots, k$, and $\hat{\eta}_l^2$ is in effect $\hat{v}_l$ the estimated variance of $\hat{\kappa}_l$ from study $l(l = 1, 2, \ldots, k)$. Sidik and Jonkman (2005) derived the simple heterogeneity variance estimator as

$$\hat{\tau}_s^2 = \frac{1}{\kappa - 1} \sum_{l=1}^{k} \hat{u}_l^{-1}(\hat{\kappa}_l - \hat{\hat{\kappa}})^2,$$ (6.18)

where

$$\hat{\hat{\kappa}} = \frac{\sum_{l=1}^{k} \hat{u}_l^{-1}\hat{\kappa}_l}{\sum_{l=1}^{k} \hat{u}_l^{-1}}, \quad \hat{u}_l = \frac{\hat{\eta}_l^2}{\hat{\tau}_o^2}.$$ (6.19)

Inference on the overall agreement coefficient κ may be conducted by constructing a weighted sample average as in the fixed effect model such that

$$\tilde{\kappa} = \frac{\sum_{l=1}^{k} \hat{W}_k \hat{\kappa}_l}{\sum_{l=1}^{k} \hat{W}_k},$$ (6.20)

where $\hat{W}_l = \left(\hat{v} + \hat{\tau}_s^2\right)^{-1}$. The variance of is $\text{var}(\tilde{\kappa}) = (\sum_{l=1}^{k}\hat{W})^{-1}$ and the $100(1 - \alpha)\%$ confidence on κ is given as $\tilde{\kappa} \pm z_{\alpha/2}\sqrt{\text{var}(\tilde{\kappa})}$.

Example 6.8

This example is for illustrative purpose. In four independent studies, two randomly assigned raters scored n_i independent subjects for the presence or absence of a certain condition. The data are summarized in Tables 6.23 and 6.24 for this special case of several 2×2 tables.

Based on the fixed model $\hat{\kappa}_c = 0.622$, $SE(\hat{\kappa}_c) = 0.1458$. Therefore, the 95% on κ_c is (0.336, 0.908).

TABLE 6.23

Data Layout of Four Studies to Assess Agreement between Two Raters on the Same Outcome

Rater a	Rater b	X_{ij}	Rater a	Rater b	X_{ij}	Rater a	Rater b	X_{ij}	Rater a	Rater b	X_{ij}
Yes	Yes	2	Yes	No	1	Yes	Yes	2	No	No	0
Yes	Yes	2	Yes	Yes	2	Yes	Yes	2	No	No	0
No	No	0	Yes	Yes	2	No	Yes	1	Yes	Yes	2
Yes	No	1	Yes	Yes	1	No	No	0	No	Yes	1
No	Yes	1	No	No	0	No	No	0	Yes	Yes	2
Yes	Yes	2				No	No	0	Yes	Yes	2
No	No	0				No	No	0			
						Yes	Yes	2			
						No	No	0			
						No	No	0			

TABLE 6.24

Summary Results for the Data in Table 6.23

c	n_1	n_2	n_3	n_4
0	2	1	6	2
1	2	2	1	1
2	3	2	3	3
n_i	7	5	10	6
$\hat{\pi}_i$	0.57	0.60	0.35	0.58
$\hat{\kappa}_i$	0.417	0.167	0.780	0.657

To construct a confidence interval on the combined estimator for kappa under the random effect model, we found $\hat{\tau}_o^2 = 0.0552$, $\hat{\theta} = 0.579$, $\hat{\tau}_s^2 = 0.0234$, and $\tilde{\kappa} = 0.5988$. Moreover, $SE(\tilde{\kappa}) = \sqrt{0.0286} = 0.169$. The 95% confidence interval on the overall coefficient of agreement based on the random effect model is (0.267, 0.930). The interval is slightly wider than the interval based on the fixed model.

EXERCISES

E6.1　For the CVM data given in Table 6.11, you have the number of ratings per subject $n = 4$. Derive the variance estimator of the measure of agreement using the GEE expression given in Equation 6.8.

E6.2　Again, for the same data table in Exercise 6.1, calculate the variance estimator for the measure of agreement using the expression in Equation 6.10.

E6.3　Noting that the two variance expressions in the above exercises are first order approximations of the variance of which the expression yields a shorter 95% confidence interval on the ICC as a measure of agreement.

E6.4　Assume that the correlations between any pair of the pairwise measures of agreements calculated from Table 6.11 are negligible. Moreover, assume that we use the simple average of the pairwise agreement ($\hat{\rho}_{ave}$, say) as an estimate of the overall agreement. Find the variance of $\hat{\rho}_{ave}$. Compare the variance of $\hat{\rho}_{ave}$ with the variances obtained from Exercises 6.1 and 6.2.

E6.5　For the data in Table 6.19, define the response variable y such that $y_{ij} = 1$ if the response of the ith subject recorded by the jth laboratory as R or BL, and $y_{ij} = 0$ if the response is NR.

　　Therefore, the situation is now reduced to multiple raters and two categories. Find the one-way and the two-way random effects estimators of the ICC, denoted respectively by $\hat{\rho}_1$ and $\hat{\rho}_2$, and compare their values to the average of the pairwise measures of agreement, and the $\hat{\kappa}_{mc}$.

E6.6 For the data in Table 6.19, calculate the pairwise measure of category distiguishability, δ as derived by Darroch and McCloud (1986). Calculate the simple average of the six measures, and compare its value to $\hat{\kappa}_{mc}$.

References

Agresti, A. 1988. A model for agreement between ratings on an ordinal scale. *Biometrics*, 44, 539–548.

Agresti, A. 1992. Modelling pattern of agreement and disagreement. *Statistical Methods in Medical Research*, 1, 201–218.

Albaum, M.N., Hill, L.C., Murphy, M., Li, Y., Fuhrman, C.R., Britton, C.A., Kapoor, W.N., and Fine, M.J. 1996. Interobserver reliability of the chest radiograph in community-acquired pneumonia. *Chest*, 110, 343–350.

Alswalmeh, Y.M. and Feldt, L.S. 1994. Testing the equality of two related intraclass reliability coefficients. *Applied Psychological Measurements*, 18, 183–190.

Altman, D.G. and Bland, J.M. 1983. Measurement in medicine: The analysis of method comparison studies. *The Statistician*, 32, 307–317.

Ashton, E., Takahashi, C., Berg, M., Goodman, A., Totterman, S., and Ekholm, S. 2003. *Accuracy and Reproducibility of Manual and Semi-Automated Quantification of MS Lesions in MRI.* Technical Report, Department of Radiology, University of Rochester Medical Center, Rochester, NY.

Asmar, L., Gehan, E., Newton, W., et al. 1994. Agreement among and within groups of pathologists in the classification of rhabdomyosarcoma and related childhood sarcomas. *Cancer*, 74(9), 2579–2588.

Bahadur, R. 1961. A representation of the joint distribution of responses to dichotomous items. In *Studies in Item Analysis and Prediction*, Solomon, H. (ed.), Palo Alto, CA: Stanford University Press, pp. 158–176.

Baker, S.G., Freedman, L.S., and Parmar, M.K.B. 1991. Using replicate observations in observer agreement studies with binary assessments. *Biometrics*, 47, 1327–1338.

Barlow, W. 1996. Measurement of interrater agreement with adjustment for covariates. *Biometrics*, 52, 695–702.

Barlow, W., Lai, M.Y., and Azen, S.P. 1991. A comparison of methods for calculating a stratified kappa. *Statistics in Medicine*, 10, 1465–1472.

Barnett, V.D. 1969. Simultaneous pair-wise linear structural relationships. *Biometrics*, 25, 129–142.

Bartfay, E. and Donner, A. 2000. The effect of collapsing multinomial data when assessing agreement. *International Journal of Epidemiology*, 29, 1070–1075.

Bartfay, E. and Donner, A. 2001. Statistical inference for interrater agreement studies with nominal outcome data. *The Statistician*, 50, 135–146.

Bartko, J. 1976. On various interclass correlation reliability coefficients. *Psychological Bulletin*, 83, 762–765.

Bartko, J.J. 1966. The intraclass correlation coefficient as a measure of reliability. *Psychological Reports*, 19, 3–11.

Bartko, J.J. 1994. General methodology II. Measures of agreement: A single procedure. *Statistics in Medicine*, 13, 737–745.

Becker, S., Al-Zaid, K., and Al-Faris, E. 2002. Screening for somatization and depression in Saudi Arabia: A validation study of the PHQ in primary care. *International Journal of Psychiatry in Medicine*, 32(3), 271–283.

Biggerstaff, B.J. and Tweedie, R.L. 1997. Incorporating variability in estimates of heterogeneity in the random effects model in meta-analysis. *Statistics in Medicine*, 16, 753–768.

Birkelo, C.C. Chamberlin, W.E., Phelps, P.S., Schools, P.E., Zacks, D., and Yerushalmy, J. 1947. Tuberculosis case-finding—A comparison of effectiveness of various roentgenographic and photofluorographic methods. *Journal of American Medical Association*, 133, 359–366.

Bishop, J., Carlin, J., and Nolan, T. 1992. Evaluation of the properties and reliability of a clinical severity scale for acute asthma in children. *Journal of Clinical Epidemiology*, 45(1), 71–76.

Bland, J.M. and Altman, D.G. 1986. Statistical methods for assessing agreement between two methods. *Lancet*, 1, 307–310.

Bland, J.M. and Altman, D.G. 1999. Measuring agreement in method comparison studies. *Statistical Methods in Medical Research*, 8, 136–160.

Bland, J.M. and Altman, D. 2007. Agreement between methods of measurement with multiple observations per individual. *Journal of Biopharmaceutical Statistics*, 17, 571–582.

Bloch, D.A. and Kraemer, H.C. 1989. 2 × 2 kappa coefficients: Measures of agreement or association. *Biometrics*, 45, 269–287.

Bonett, D.G. 2002. Sample size requirements for estimating intraclass correlations with desired precision. *Statistics in Medicine*, 21, 1331–1335.

Bowerman, G.P., Markman, S., Thompson, G., Minuk, T., Chirawatkul, A., and Roberts, R.S. 1990. Assessment of observer variation in measuring the radiographic vertebral index in patients with multiple myeloma. *Journal of Clinical Epidemiology*, 43, 833–840.

Bradley, E.L. and Blackwood, L.G. 1989. Comparing paired data: A simultaneous test for means and variances. *The American Statistician*, 43(4), 234–235.

Broemeling, L.D. 2009. *Bayesian Methods for Measures of Agreement*. Boca Raton, FL: Chapman & Hall/CRC Press.

Byrt, T., Bishop, J., and Carlin, J.B. 1993. Bias, prevalence, and kappa. *Journal of Clinical Epidemiology*, 46, 423–429.

Cappelleri, J.C. and Ting, N. 2003. A modified large sample approach to approximate interval estimation for a particular intraclass correlation coefficient. *Statistics in Medicine*, 22(1), 1861–1877.

Carrasco, J.L. and Jover, J.L. 2003. The concordance correlation coefficient estimated through variance components. *IX Conferencia Española de Biometria*, La Coruña, 28–30 de mayo; 1–4.

Carroll, R.L. and Ruppert, D. 1996. The use and misuse of orthogonal regression in linear error-in-variables models. *The American Statistician*, 50, 1–6.

Carroll, R.L., Ruppert, D., and Stafanski, L. 1995. *Measurement Error in Non-Linear Models*. New York, NY: Chapman & Hall.

Carstensen, B., Simpson, J., and Gurrin, L.C. 2008. Statistical models for assessing agreement in method comparison studies with replicate measurements. *The International Journal of Biostatistics*, 14(1), Article 16.

Chow, S.-C. and Tse, S.-K. 1990. A related problem in bioavailability/bioequivalence studies—Estimation of the intrasubject variability with common CV. *Biometrical Journal*, 32, 597–607.

Cicchetti, D.V. and Fleiss, J.L. 1977. Comparison of the null distributions of weighted kappa and the C ordinal statistic. *Applied Psychological Measurements*, 1, 195–201.

Cicchetti, D.V. and Feinstein, A.R. 1990. High agreement but low kappa. I. The problem of two paradoxes. *Journal of Clinical Epidemiology*, 6, 543–549.

Clarke, E.A. and Anderson, T.W. 1979. Does screening by "Pap" smears help prevent cervical cancer? *The Lancet*, 7, 1–4.

Clarke, E.A., Hatcher, J., and McKeown-Eyssen, G.E. 1985. Cervical dysplasia: Association with sexual behavior, smoking, and oral contraceptive use. *American Journal of Obstetrics and Gynecology*, 151, 612–616.

Cochrane, A.L., Chapman, P.J., and Oldham, P.D. 1951. Observers' errors in taking medical histories. *The Lancet*, 257, 1007–1009.

Cohen, J. 1960. A coefficient of agreement for nominal scales. *Educational and Psychological Measurement*, 20, 37–46.

Cohen, J. 1968. Weighted kappa: Nominal scale agreement with provision for scaled disagreement or partial credit. *Psychological Bulletin*, 70, 213–220.

Coughlin, S.S., Pickle, L.W., Goodman, M.T., and Wilkens, L.R. 1992. The logistic modeling of interobserver agreement. *Journal of Clinical Epidemiology*, 45(11), 1237–1241.

Cox, D.R. and Hinkely, D.V. 1974. *Theoretical Statistics*. London: Chapman & Hall.

Cox, D.R. and Snell, E.J. 1989. *Analysis of Binary Data*, 2nd edn., London: Chapman & Hall.

Cramer, H. 1946. *Mathematical Methods of Statistics*. NJ: Princeton University Press.

Cronbach, L.J., Nageswari, R., and Gleser, G.C. 1963. Theory of generalizability: A liberation of reliability theory. *The British Journal of Statistical Psychology*, 16, 137–163.

Crowder, M. 1978. Beta-binomial ANOVA for proportions. *Applied Statistics*, 27, 34–37.

Darroch, J.N. and McCloud, P.I. 1986. Category distinguishability and observer agreement. *Australian Journal of Statistics*, 28(3), 371–388.

Davies, L.G. 1957. Observer variation in reports on electrocardiograms. *British Medical Association*, XI, 153–161.

Davies, M. and Fleiss, J.L. 1982. Measuring agreement for multinomial data. *Biometrics*, 38, 1048–1051.

Deming, W.E. 1943. *Statistical Adjustment of Data*. New York, NY: Wiley.

Dhingsa, R., Finlay, D.B., Robinson, G.D., and Liddicoat, A.J. 2002. Assessment of agreement between general practitioners and radiologists as to whether a radiation exposure is justified. *The British Journal of Radiology*, 75, 136–139.

Diaz, L.K., Sahin, A., and Sneige, N. 2004. Interobserver agreement for estrogen receptor immuno histochemical analysis in breast cancer: A comparison of manual and computer-assisted scoring methods. *Annals of Diagnostic Pathology*, 8(1), 23–27.

Donner, A. 1986. A review of inference procedures for the intraclass correlation coefficient in the one-way random effects model. *International Statistical Review*, 54(1), 67–82.

Donner, A. and Bull, S. 1983. Inferences concerning a common intraclass correlation coefficient. *Biometrics*, 39, 771–775.

Donner, A. and Wells, G. 1986. A comparison of confidence interval methods for the intraclass correlation coefficient. *Biometrics*, 42, 401–412.

Donner, A. and Eliasziw, M. 1987. Sample size requirements for reliability studies. *Statistics in Medicine*, 6, 441–448.

Donner, A. and Eliasziw, M. 1992. A goodness-of-fit approach to inference procedures for the kappa statistic: Confidence interval construction, significance-testing and sample size estimation. *Statistics in Medicine*, 11, 1511–1519.

Donner, A., Eliasziw, M., and Klar, N. 1996. Testing the homogeneity of kappa statistics. *Biometrics*, 52(1), 176–183.

Donner, A., Shoukri, M.M., Klar, N., and Bartfay, E. 2000. Testing the quality of two independent kappa statistics. *Statistics in Medicine*, 19, 373–387.

Donner, A. and Zou, G. 2002. Testing the equality of dependent intraclass correlation coefficients. *The Statistician*, 51(Part 3), 367–379.

Dunn, G. 1992. Design and analysis of reliability studies. *Statistical Methods in Medical Research*, 1, 123–157.

Ebel, R.L. 1951. Estimation of the reliability of ratings. *Psychometrika*, 16, 407–424.

Edwards, A.L. 1948. Note on the correction for continuity in testing the significance of the difference between correlated proportions. *Psychometrika*, 13, 185–187.

Eliasziw, M. and Donner, A. 1987. A cost-function approach to the design of reliability studies. *Statistics in Medicine*, 6, 647–655.

Elston, R. 1975. On the correlation between correlations. *Biometrika*, 62(1), 133–140.

Fanshawe, T., Lynch, A., Ellis, I., Green, A., and Honka, R. 2008. Assessing agreement between multiple raters with missing rating information, applied to breast cancer tumor grading. *PLOS ONE* 3(8), e2925.

Fawzy, M.E., Mercer, E.N., Dunn, B., Al-Amri M., and Andaya, W. 1989. Doppler echocardiography in the evaluation of tricuspid stenosis. *European Heart Journal*, 10, 985–990.

Feinstein, A.R. and Cicchetti, D.V. 1990. High agreement but low kappa: I. The problems of two paradoxes. *Journal of Clinical Epidemiology*, 43, 543–548.

Feinstein, A.R. and Cicchetti, D.V. 1990. High agreement but low kappa II. Resolving the paradoxes. *Journal of Clinical Epidemiology*, 43, 551–558.

Feltz, C.J. and Miller, G.E. 1996. An asymptotic test for the coefficients of variation from k populations. *Statistics in Medicine*, 15, 647–658.

Fisher, R.A. 1925. *Statistical Methods for Research, Workers*, 1st edn. Edinburgh & London: Oliver & Boyd.

Fisher, R.A. 1958. *Statistical Methods for Research Workers*. New York, NY: Hafner.

Fitzmaurice, G.M., Laird, N.M., Zahner, G.E.P., and Daskalakis, C. 1995. Bivariate logistic regression analysis of childhood psychopathology ratings using multiple informants. *American Journal of Epidemiology*, 142, 1194–1203.

Fleiss, J.L. 1966. Assessing the accuracy of multivariate observations. *Journal of the American Statistical Association*, 61, 403–412.

Fleiss, J.L. 1971. Measuring nominal scale agreement among many raters. *Psychological Bulletin*, 76, 378–382.

Fleiss, J.L. 1981. *Statistical Methods for Rates and Proportions*, New York, NY: John Wiley and Sons.

Fleiss, J.L. 1986. *The Design and Analysis of Clinical Experiment*. New York, NY: Wiley.

Fleiss, J.L. and Everitt, B.S. 1971. Comparing the marginal totals of square contingency tables. *British Journal of Mathematical and Statistical Psychology*, 24, 117–123.

Fleiss, J.L. and Cohen, J. 1973. The equivalence of the weighted kappa and the intraclass correlation coefficient as a measure of reliability. *Educational and Psychological Measurements*, 33, 613–619.

Fleiss, J.L. and Cicchetti, D.V. 1978. Inferences about weighted kappa in the nonnull case. *Applied Psychological Measurements*, 2, 113–117.

Fleiss, J.L. and Shrout P. 1978. Approximate interval estimation for a certain intraclass correlation coefficient. *Psychometrika* 43, 259–262.

Fleiss, J.L. and Cuzick, J. 1979. The reliability of dichotomous judgments: Unequal number of judgments per subject. *Applied Psychological Measurement*, 3, 537–542.

Fleiss, J.L., Cohen, J., and Everitt, B.S. 1969. Large sample standard errors of kappa and weighted kappa. *Psychological Bulletin*, 72, 323–327.

Fletcher, C.M. 1952. The clinical diagnosis of pulmonary emphysema—an experimental study. *Proceedings of the Royal Society of Medicine*, 45, 577–584.

Fletcher, C.M. and Oldham, P.D. 1964. Bibliography on observer error and variation. In *Medical Surveys and Clinical Trials*, 2nd edn. Witts, L.J. (ed.), New York, NY: Oxford University Press, pp. 39–44.

Flynn, N.T., Whitley, E., and Peters, T. 2002. Recruitment strategy in a cluster randomized trial–cost implications. *Statistics in Medicine*, 21, 397–405.

Fuller, W. 1987. *Measurement Errors Models*. New York, NY: Wiley.

Fung, W.K. and Tsong, T.S. 1998. A simulation study comparing tests for the equality of coefficients of variation. *Statistics in Medicine*, 17, 2003–2014.

Gastwirth, J. 1987. The statistical precision of medical screening test. *Statistical Sciences*, 2, 213–238.

Giraudeau, B. and Mary, J.Y. 2001. Planning a reproducibility study: How many subjects and how many replicates per subject for an expected width of the 95 per cent confidence interval of the intraclass correlation coefficient. *Statistics in Medicine*, 20, 3205–3214.

Gong, G. and Samaniego, F. 1981. Pseudo maximum likelihood estimation: Theory and applications. *Annals of Statistics* 9, 861–869.

Grubbs, F.E. 1948. On estimating precision of measuring instruments and product variability. *Journal of the American Statistical Association*, 43, 243–264.

Gupta, R.C. and Ma, S. 1996. Testing the equality of coefficients of variation from normal populations. *Communications in Statistics–Theory and Methods*, 25, 115–132.

Guyatt, G., Walter, S., and Norman, G. 1987. Measuring change over time: Assessing the usefulness of evaluative instruments. *Journal of Chronic Diseases*, 40, 171–178.

Haber, M. and Barnhart, H. 2006. Coefficient of agreement for fixed observers. *Statistical Methods in Medical Research*, 15, 255–271.

Haggard, E.A. 1958. *Intraclass Correlation and the Analysis of Variance*. New York, NY: Dryden Press.

Hale, C.A. and Fleiss, J.L. 1993. Interval estimation under two study designs for kappa with binary classifications. *Biometrics*, 49(2), 523–534.

Hand, D.J. 1996. Statistics and the theory of measurements (with discussion). *Journal of the Royal Statistical Society, Series A*, 159, 445–492.

Hannah, M.C., Hopper, J.L., and Mathews, J.D. 1983. Twin concordance for a binary trait. I Statistical models illustrated with data on drinking status. *Acta geneticae medicae et gemellologiae*, 32, 127–137.

Hardy, R.J. and Thompson, S.G. 1998. Detecting and describing heterogeneity in meta-analysis. *Statistics in Medicine*, 17, 84–856.

Haseman, J.K. and Kupper, L.L. 1979. Analysis of dichotomous response data from certain toxicological experiments. *Biometrics*, 35, 281–293.

Hemmersley, I.M. 1949. The unbiased estimate and standard error of the intraclass variance. *Metron*, 15, 189–205.

Hirji, K.F. and Rosove, M.H. 1990. A note on interrater agreement. *Statistics in Medicine*, 9, 835–839.

Hoehler, F.K. 2000. Bias and prevalence effects on kappa viewed in terms of sensitivity and specificity. *Journal of Clinical Epidemiology*, 53, 499–503.

Holley, J.W. and Guildford, J.P. 1964. A note on the G index of agreement. *Educational and Psychological Measures*, 32, 281–288.

Hollis, S. 1996. Analysis of method comparison studies (Editorial). *Annals of Clinical Biochemistry*, 33, 1–4.

Hosmer, D.W. and Lemeshow, S. 1989. *Applied Logistic Regression*. New York, NY: Wiley and Sons.

Hui, S.L. and Walter, S.D. 1980. Estimating the error rates of diagnostic tests. *Biometrics*, 36, 167–171.

Jaech, J.L. 1985. *Statistical Analysis of Measurement Errors*. New York, NY: Wiley.

Jung, J.A., Coakley, F.V., Vigneron, D., et al. 2004. Prostate depiction at endorectal MR spectroscopic imaging: Investigation of a standardized evaluation system. *Radiology*, 233, 701–708.

Kendall, M. and Ord, K. 1989. *Advanced Theory of Statistics*, Vol. I. London: Griffin.

Kendall, M. and Ord, K. 1991. *Advanced Theory of Statistics: Classical Inference and the Linear Model*, Vol. II, London: Griffin.

Kirchner, H.L. and Lemke, J.H. 2002. Simultaneous estimation of interrater and intrarater agreement for multiple raters under order restriction for a binary trait. *Statistics in Medicine*, 21, 1761–1772.

Klar, N., Lipsitz, S.R., and Ibrahim, J. 2000. An estimating equations approach for modeling kappa. *Biometrical Journal*, 42, 45–58.

Koch, D.D. and Peters, T. 1999. Selection and evaluation of methods. In *Tietz Textbook of Clinical Chemistry*, 3rd edn. Burts, C.A. and Ashwood, E.R. (eds.), Philadelphia: W.B. Saunders Company, pp. 320–335.

Konishi, S. and Gupta, A. 1989. Testing the equality of several intraclass correlation coefficients. *Journal of Statistical Planning and Inference*, 21(1), 93–105.

Kraemer, H.C. 1979. Ramification of a population model for kappa as a coefficient of reliability. *Psychometrika*, 44, 461–472.

Kreppendorff, K. 1970. Bivariate agreement coefficient for reliability data. In *Sociological Methodology*. Borgatta, F. and Bohrnsteadt, G.W. (eds.), San Francisco: Jossey-Bass, pp. 139–150.

Kummel, C.H. 1879. Reduction of observed equations which contain more than one observed quantity. *The Analyst*, 6, 97–105.

Kundel, H.L. and Polansky, M. 2003. Measurement of observer agreement. *Radiology*, 228, 303–308.

Lancester, H.O. 1969. *The Chi-Squared Distributions*. New York, NY: Wiley.

Landis, J.R. and Koch, G.G. 1977a. The measurement of observer agreement for categorical data. *Biometrics*, 33, 159–174.

Landis, J.R. and Koch, G.G. 1977b. A one-way components of variance model for categorical data. *Biometrics*, 33, 671–679.

Lehmann, E.L. 1996. *Testing Statistical Hypotheses*, 2nd Edn. New York, NY: Wiley.

Liang, K.Y. and Zeger, S.L. 1986. Longitudinal data analysis using generalized linear models. *Biometrika*, 73, 13–22.

Liehr, P., Dedo, Y.L., Torres, S., and Meininger, J.C. 1995. Assessing agreement between clinical measurement methods. *Heart and Lung*, 24(3), 240–245.

Lin, L.I. 1989. A concordance correlation coefficient to evaluate reproducibility. *Biometrics*, 45, 255–268.

Linnett, K. 1990. Estimation of the linear relationship between the measurements of two methods with proportional errors. *Statistics in Medicine*, 9, 1463–1473.

Linnett, K. 1999. Necessary sample size for method comparison studies based on regression analysis. *Clinical Chemistry*, 45(6), 882–894.

Lipstiz, S.R., Laird, N.M., and Brennan, T.A. 1994. Simple moment estimates of the *k*-coefficient and its variance. *Applied Statistics*, 43(2), 309–323.

Mak, H.K., Yau, K.K., and Chan, B.P. 2004. Prevalence-adjusted bias adjusted kappa values as additional indicators to measure observer agreement: (Letter to the Editor). *Radiology*, 232(1), 302–303.

Mak, T.K. 1988 Analyzing intraclass correlation for dichotomous variables. *Applied Statistics*, 20, 37–46.

Maloney, C.J. and Rastogi, S.C. 1970. Significance test for Grubbs's estimators. *Biometrics*, 26, 671–676.

Maxwell, A.E. 1977. Coefficient of agreement between observers and their interpretation. *British Journal of Psychiatry*, 130, 79–83.

McGraw, K.O. and Wong, S.P. 1966. Forming inferences about some intraclass correlation coefficients. *Psychological Methods*, 1(1), 30–46.

McKenzie, D.P., Mackinnan, A.J., Peladeau, N., Onghena, P., Bruce, P.C., Clarke, D.M., Harrigan, S., and McGorry, P.D. 1996. Comparing correlated kappas by resampling: Is one level of agreement significantly different from another. *Journal of Psychiatry Research*, 30, 483–492.

McNemar, Q. 1947. Note on the sampling error of the difference between correlated proportions or percentages. *Psychometrika*, 12, 153–157.

Mekibib, A., Donner, A., and Eliasziw, M. 2001. A general goodness-of-fit approach for inference procedures concerning the kappa statistic. *Statistics in Medicine*, 20, 2479–2488.

Mian, I.U.H. and Shoukri, M.M. 1997. Statistical analysis of intraclass correlation from multiple samples with applications to arterial blood pressure data. *Statistics in Medicine*, 16(13), 1497–1514.

Miller, G.E. 1991a. Asymptotic test statistic for coefficients of variation. *Communication in Statistics–Theory and Methods*, 20, 3351–3363.

Miller, G.E. 1991b. Use of the squared ranks test to test for the equality of the coefficients of variation. *Communications in Statistics–Simulation and Computation*, 20, 743–750.

Molenberghs, G., Fitzmaurice, G.M., and Lipsitz, S.R. 1996. Efficient estimation of the intraclass correlation for a binary trait. *Journal of Agricultural, Biological, and Environmental Statistics*, 1, 78–96.

Morgan, W. 1939. A test for the significance of the difference between two variances in a sample from bivariate population. *Biometrika*, 31, 13–19.

Muller, R. and Butter, P. 1994. A critical discussion of intraclass correlation coefficients. *Statistics in Medicine*, 13, 2465–2476.

Neyman, J. 1959. Optimal asymptotic tests of composite hypotheses. In *Probability and Statistics: The Harold Cramer Volume*. Grenander, V. (ed.), New York, NY: Wiley, pp. 213–234.

Neyman, J. and Pearson, E.S. 1928. On the use and interpretation of certain test criteria for purposes of statistical inferences. *Biometrika*, 20A, 176–240, 263–294.

Oden, N. 1991. Estimating kappa from binocular data. *Statistics in Medicine*, 10, 1303–1311.

Paul, S.R. and Barnwal, R.K. 1990. Maximum likelihood estimation and a C(∞) test for a common intraclass correlation. *The Statistician*, 39, 19–24.

Pitman, E.J.G. 1939. A note on normal correlation. *Biometrika*, 31, 9–12.

Powell, K.A., Obouchowski, N.A., Chilcote, W.A., Barry, M.M., Ganobcik, S.N., and Cardenosa, G. 1999. File-screen versus digitized mammography: Assessment of clinical equivalence. *American Journal of Roentgenology*, 173, 889–894.

Prentice, R. 1988. Correlated binary regression with covariates specific to each binary observation. *Biometrics*, 44, 1033–1048.

Quan, H. and Shih, W. 1996 Assessing reproducibility by the within-subject coefficient of variation with random effects models. *Biometrics*, 52, 1195–1203.

Rajaratnam, M. 1960. Reliability formulas for independent decision data when reliability data are matched. *Psychometrika*, 25, 261–271.

Ramasundarahettige, C.F., Donner, A., and Zou, G.Y. 2009. Confidence interval construction for the difference between two dependent intraclass correlation coefficients. *Statistics in Medicine*, 29, 1041–1053.

Rao, S.S. 1984. *Optimization: Theory and Applications*, 2nd edn., New Delhi: Wiley Eastern Limited.

Rifkin, M.D., Zerhouni, E.A., Constantine, M.D., Gastonis, C.A., Quint, L.E., Paushter, D.M., Epstein, J.I., Hamper, U., Walsh, P.C., and McNeil, B.J. 1990. Comparison of magnetic resonance imaging and ultrasonography in staging early prostate cancer. *The New England Journal of Medicine*, 323(10), 621–626.

Rogan, W. and Gladen, B. 1978. Estimating prevalence from the results of a screening test. *American Journal of Epidemiology*, 107, 71–76.

Rosner, B. 1989. Multivariate methods for clustered binary data with more than one level of nesting. *Journal of the American Statistical Association*, 84, 373–380.

Rosner, B. 1992. Multivariate methods for clustered binary data with multiple subclasses, with application to binary longitudinal data. *Biometrics*, 48, 721–731.

Rousson, V., Gasser, T., and Seifert, B. 2002. Assessing intrarater, interrater, and test–retest reliability of continuous measurements. *Statistics in Medicine*, 21, 3431–3446.

Saito, Y., Sozu, T., Hamada, C., and Yoshimura, I. 2006. Effective number of subjects and number of replicates of raters for interrater reliability studies. *Statistics in Medicine*, 25, 1547–1560.

Schaalje, B.G. and Butts, R.A. 1993. Some effects of ignoring correlated measurements errors in straight line regression and prediction. *Biometrics*, 49, 1262–1267.

Scheffe, H. 1959. *Analysis of Variance.* New York, NY: Wiley.

Schouten, H.J.A. 1993. Estimating kappa from binocular data and comparing marginal probabilities. *Statistics in Medicine*, 12, 2207–2217.

Schwartz, L.H., Ginsberg, M.S., DeCorato, D., Rothenberg, L.N., Einstein, S., Kijweski, P., and Panicek, D. 2000. Evaluation of tumor measurements in oncology: Use of film-based and electronic techniques. *Journal of Clinical Oncology*, 18(10), 2179–2184.

Scott, W.A. 1955. Reliability of content analysis: The case of nominal scale coding. *Public Opinion Quarterly*, 19, 321–325.

Searle, R.S., Casella, G., and McCulloch, C.E. 1992. *Variance Components.* New York, NY: Wiley-Interscience.

Serfontein, G.L. and Jaroszewitz, A.M. 1978. Estimation of gestational age at birth—Comparison of two methods. *Archives of Diseases in Childhood* 53, 509–511.

Shavelson, R.J., Rowley, G.L., and Webb, N.M. 1989. Generalizability theory. *American Psychology*, 44, 922–932.

Shoukri, M., El-Kum, N., and Walter, S.D. 2006. Interval estimation and optimal design for the within-subject coefficient of variation for continuous and binary variables. *BMC Medical Research Methodology* 6, 24.

Shoukri, M.M. 1999. Agreement. In *Encyclopedia of Biostatistics*. Armitage, P. and Colton, T. (eds.), New York, NY: John Wiley and Sons.

Shoukri, M.M. 2000. Agreement. In *Encyclopedia of Epidemiology*. Gail, M. (ed.), New York, NY: John Wiley and Sons.

Shoukri, M.M., Martin, S.W., and Mian, I.U.H. 1995. Maximum likelihood estimation of the kappa coefficient from models of matched binary responses. *Statistics in Medicine*, 4, 83–99.

Shoukri, M.M. and Mian, I.U.M. 1996. Maximum likelihood estimation of the kappa coefficient from bivariate logistic regression. *Statistics in Medicine*, 15, 1409–1419.

Shoukri, M.M. and Pause, C.A. 1999. *Statistical Methods for Health Sciences*, 2nd edn., Boca Raton, FL: CRC Press.

Shoukri, M.M. and Donner, A. 2001. Efficiency considerations in the analysis of inter-observer agreement. *Biostatistics*, 2(3), 323–336.

Shoukri, M.M., Asyali, M.H., and Walter, S.W. 2003. Issues of cost and efficiency in the design of reliability studies. *Biometrics*, 59(4), 1109–1114.

Shoukri, M.M., Colak, D., Kaya, N., and Donner, A. 2008. Comparison of two dependent within subject coefficient of variation to evaluate the reproducibility of measurement devices. *BMC Medical Research Methodology*, 8(24), 1–11.

Shoukri, M.M. and Donner, A. 2009. Bivariate modeling of inter-observer agreement coefficient. *Statistics in Medicine*, 28, 430–440.

Shrout, P.E. and Fleiss, J.L. 1979. Intraclass correlations: Uses in assessing rater reliability. *Psychological Bulletin*, 86(2), 420–428.

Shrout, P.E., Spitzer, R.L., and Fleiss, J.L. 1981. Quantification of agreement in psychiatric diagnosis revisited. *Archives of General Psychiatry*, 44, 172–177.

Shukla, G.K. 1973. Some exact tests on hypothesis about Grubbs' estimators. *Biometrics*, 29, 373–377.

Sidik, K. and Jonkman, J.N. 2005. Simple heterogeneity variance estimator for meta-analysis. *Applied Statistics*, 54, 367–384.

Singhal, R.A. 1984. Effect of non-normality on the estimation of functions of variance components. *Journal of the Indian Society of Agricultural Statistics*, 35(1), 89–98.

Sisson, H.A. 1975. Agreement and disagreement between pathologists in histological diagnosis. *Postgraduate Medical Journal*, 51, 685–698.

Smith, C.A.B. 1956. On the estimation of intraclass correlation. *Ann. Hum. Genet.* 21, 363–373.

Snedecor, G. and Cochran, W.G. 1980. *Statistical Methods* 8th edn., Ames, IA: Iowa State University Press.

Solm, L.P., Jeroar, J., Vliagen, H.W., and Langerk, S.E. 2004. Functional significance of stenosis in coronary artery bypass grafts: Evaluation by single-photon emission computed tomography perfusion imaging, cardiovascular magnetic resonance, and angiography. *Journal of the American College of Cardiology*, 44(9), 1877–1882.

Spearman, C. 1910. Correlation calculated from faulty data. *British Journal of Psychology*, 3, 271–295.

St. Laurent, R.T. 1998. Evaluating agreement with a gold standard in method comparison studies. *Biometrics*, 54, 537–545.

Stevens, S.S. 1946. On the theory of scales of measurements. *Science*, 103, 677–680.

Stockl, D., Dewitte, K., and Theinpont, L.M. 1998. Validity of linear regression in method comparison studies: Is it limited by the statistical model or the quality of the analytical input data? *Clinical Chemistry*, 44(11), 2340–2346.

Strike, P.W. 1995. *Measurements in Laboratory Medicine*. Oxford: Butterworth.

Stuart, A. and Ord, K. 1987. *Kendall's Advanced Theory of Statistics*, Vol. 1, 5th edn., London: Griffin, p. 324.

Sukhatme, P.V., Sukhatme, B.V., Sukhatme, S., and Asok, C. 1984. *Sampling Theory of Surveys with Applications*. Ames, IA: Iowa State University Press.

Tanner, M. and Young, M.A. 1985. Modeling agreement among raters. *Journal of the American Statistical Association*, 80, 175–180.

Tenenbein, A. 1970. A double sampling scheme for estimating data with misclassifications. *Journal of the American Statistical Association*, 65, 1350–1361.

Thompson, W.A., Jr. 1962. The problem of negative estimates of variance components. *Annals of Mathematical Statistics*, 33, 273–289.

Thompson, W.D. and Walter, S.D. 1988. A reappraisal of the kappa coefficient. *Journal of Clinical Epidemiology*, 41(10), 949–958.

Tian, L. 2005. Inference on the common coefficient of variation. *Statistics in Medicine*, 24, 213–2220.

Tukey, J.W. 1956. Variance of variance components. I. Balanced designs. *Annals of Mathematical Statistics*, 27, 722–736.

Vach, W. 2005. The dependence of Cohen's kappa on the prevalence does not matter. *Journal of Clinical Epidemiology*, 58(7), 655–661.

Vangel, M.G. 1996. Confidence intervals for a normal coefficient of variation. *American Statistician*, 50, 21–26.

Von Eye, A. and Mun, E.Y. 2005. *Analyzing Rater Agreement, Manifest Variable Methods*. Manwash, NJ, London: Lawrence Erlbaum Associates.

Walter, D.S., Eliasziw, M., and Donner, A. 1998. Sample size and optimal design for reliability studies. *Statistics in Medicine*, 17, 101–110.

Walter, S.D., Clarke, E.A., Hatcher, J., and Stitt, L.W. 1988. A comparison of physician and patient reports of Pap smear and histories. *Journal of Clinical Epidemiology*, 41(4), 401–410.

Wax, P., Hoffman, S., and Goldfrank, L.R. 1992. Rapid quantitative determination of blood alcohol concentration in the emergency department using an electrochemical method. *Annals of Emergency Medicine*, 21(3), 254–259.

Weinberg, R. and Patel, Y.C. 1981. Simulated intraclass correlation coefficients and their z-transforms. *Journal of Statistical Computations and Simulations*, 13(1), 13–26.

Westgard, J.O. 1998. Points of care in using statistics in method-comparison studies. *Clinical Chemistry*, 44(1), 2240–2242.

Westgard, J.O. and Hunt, M.R. 1973. Use and interpretation of common statistical tests in method—Comparison studies. *Clinical Chemistry*, 19, 49–57.

Westlund, K.B. and Kurland, L.T. 1953. Studies on multiple sclerosis in Winnepeg, Manitoba and New Orleans, Louisiana. *American Journal of Hygiene*, 57, 380–396.

Williams, G.W. 1976. Comparing the joint agreement of several raters with another rater. *Biometrics*, 32, 619–627.

Williamson, J.M. and Manatunga, A.K. 1997. The consultant's forum: Assessing inter-rater agreement from dependent data. *Biometrics*, 53, 707–714.

Woolson, R.F. 1987. *Statistical Methods for the Analysis of Biomedical Data*. New York, NY: Wiley & Sons.

Yerushalmy, J., Harkness, J.T., Cope, J.N., and Kennedy, B.R. 1950. The role of dual reading in mass radiography. *American Review of Tuberculosis*, 61, 443–464.

Zou, K.H. and McDermott, M.P. 1999. Higher-moment approaches to approximate interval estimation for a certain intraclass correlation coefficient. *Statistics in Medicine*, 18, 2051–2061.

Index

Milton Keynes UK
Ingram Content Group UK Ltd.
UKHW040445071024
449327UK00020B/1016